INDEPENDENT LEARNING PROJECT FOR ADVANCED CHEMISTRY

ILPAC
second edition

10

ORGANIC

BIG MOLECULES

REVISED BY ANN LAINCHBURY JOHN STEPHENS ALEC THOMPSON

JOHN MURRAY

■ ACKNOWLEDGEMENTS

We are grateful to CLEAPSS/ASE Laboratory Standards Committee for ensuring that the text meets with current safety recommendations.

Thanks are due to the following examination boards for permission to reproduce questions from past A-level papers:

Associated Examining Board: Exercises 32, p. 37 (1992); 101, p. 112 (1990). Joint Matriculation Board: Exercise 40, p. 48 (1991); Teacher-marked exercise, p. 101 (1978). Northern Examinations and Assessment Board: Exercise 12, p. 15 (1993); Teacher-marked exercise 2, p. 55 (1993). Oxford and Cambridge Schools Examination Board: Teacher-marked exercise, p. 13 (1993); Exercises 32, p. 37 (1990); 89, p. 87 (1993); 115, p. 127 (1994); End-of-unit test Section A 6, p. 142 (1993). University of Cambridge Local Examinations Syndicate: Exercises 7, p. 9 (1979); 33, p. 38 (1993); 42, p. 51 (1994); 100, p. 112 (1979); 105, p. 117 (1996/97 specimen); 112, p. 122 (1992); 118, p. 135, 119, p. 136, 126, p. 136 (1994 specimen); 121, p. 136 (1995), 124, p. 138 (1993); Teacher-marked exercises, p. 62 (1994 specimen); 4, p. 103 (1993); p. 123 (1992); End-of-unit test Section A 1, p. 139, 2, p. 140 (1994 specimen). University of London Examinations and Assessment Council: Teacher-marked exercises, p. 65 (N 1980); part 1, p. 102 (N 1982); 1, p. 55 (N 1993); Exercises 69, p. 80 (1974); 70, p. 81 (N 1976); 80, p. 90 (1981); 86, p. 94 (N 1993); 94, p. 107 (N 1993); 99, p. 111 (1974); 100, p. 112 (1982); 102, p. 113 (N 1979); 109, p. 121 (N 1975); 110, p. 121 (N 1974); End-of-unit test Section A 3, p. 141 (N 1996 specimen); 4, p. 141 (1994); 5, p. 142 (1993); Section B 1, p. 144, 2, p. 144 (N 1996 specimen). University of Oxford Delegacy of Local Examinations: Exercise 58, p. 64 (1977); Teacher-marked exercises, p. 90 (1980); part 2, p. 103 (1981); part 3, p. 103 (1980). Welsh Joint Education Committee: Exercises 19, p. 26 (1976); 25, p. 29 (1976); 56, p. 63 (1977 & 1979); 67, p. 75 (1979); Teacher-marked exercise, p. 39 (1991).

(The examination boards accept no responsibility whatsoever for the accuracy or method of working in the answers given.)

Thanks are also due to the following: Mr Paul Hart (Unilever) for extra data for Table 7. University of Cambridge Local Examinations Syndicate for Table 19, page 126.

Photographs reproduced with kind permission of John Townson/Creation (pp. 25, 54 top, 56); Simon Fraser/Science Photo Library (p. 50); Dr Jeremy Burgess/Science Photo Library (p. 54 bottom); DRCT/Custom Medical Stock Photo (p. 61); Dr Gopal Murti/Science Photo Library (p. 67); The Image Bank (p. 87); Heini Schneebel/Science Photo Library (p. 95); Zeneca BioProducts (p. 107); CNRI/Science Photo Library (p. 128); Science Source/Science Photo Library (p. 157 left); Dr MB Hursthouse/Science Photo Library (p. 157 right). All other photographs by Last Resort Picture Library. The assistance provided by the staff and students of Djanogly City Technology College, Nottingham for the photographs of experiments is gratefully acknowledged.

The publishers have made every effort to trace copyright holders, but if they have inadvertently overlooked any they will be pleased to make the necessary arrangements at the earliest opportunity.

Original material produced by the Independent Learning Project for Advanced Chemistry sponsored by the Inner London Education Authority

First edition published in 1983
by John Murray (Publishers) Ltd
50 Albemarle Street
London W1X 4BD

Second edition 1996

British Library Cataloguing in Publication Data
A catalogue record for this book is available from the British Library.

ISBN 0–7195–5340–7

Design and layouts by John Townson/Creation.
Illustrations by Barking Dog Art.

Typeset in 10/12 pt Times and Helvetica by Wearset, Boldon, Tyne and Wear.

Printed in Great Britain by St Edmundsbury Press Ltd, Bury St Edmunds.

CONTENTS

■ BIG MOLECULES

■ Symbols used in ILPAC

 Computer program

 A-level question

 Discussion

 A-level part question

 Experiment

 A-level question;
Special Paper

 Model-making

 A-level supplementary
question

 Reading

 Revealing Exercise

 Video programme

■ International hazard symbols

 Corrosive

 Oxidising

 Explosive

 Radioactive

 Harmful or irritant

 Toxic

 Highly flammable

BIG MOLECULES

INTRODUCTION

This is the last of three volumes on organic chemistry. As before, we have not split the material into two levels of difficulty because this would cut across the more natural divisions. It is divided into eight chapters. The first six deal with carboxylic acids, carboxylic acid derivatives, lipids, amino acids, proteins and synthetic polymers. Although carboxylic acids are not big molecules we include them in this book as a logical step to amino acids and proteins. In Chapter 7, we look at the instrumental techniques of mass spectrometry, infra-red and nuclear magnetic resonance spectroscopy for the determination of structure and identification of organic compounds. The techniques of chromatography and electrophoresis are included in the sections on amino acids and proteins as well as DNA fingerprinting. Chapter 8 gives tips on how to answer examination questions.

In Appendix 4 we consider high-resolution NMR spectroscopy, that is, 'spin-spin splitting' and outline the theory of another technique, X-ray crystallography, in Appendix 5. A section on nucleic acids (including DNA) is also provided in Appendix 2.

There are nine experiments in this book (one of which is in Appendix 1).

 There are four ILPAC video programmes – 'Organic techniques I and II', 'Identifying unknown substances' and 'Instrumental techniques' – designed to accompany this book. The Royal Society of Chemistry and GlaxoWellcome have produced a video to support the teaching of instrumental techniques at the post-16 level. Your teacher may have a copy of this useful video called 'Modern chemical techniques'. They are not essential, but you should try to see them at the appropriate time if they are available.

■ Pre-knowledge

Before you start on this book you should have studied ILPAC 5, Introduction to Organic Chemistry, and ILPAC 8, Functional Groups, and also should be able to:

- ■ explain the term 'hard water';
- ■ describe the effect of hard water on soap;
- ■ write structural formulae for two complex ions of copper;
- ■ describe the action of a buffer solution;
- ■ derive and use the expression:

$$pH = pK_a - \log \frac{[acid]}{[base]}$$

■ Pre-test

To find out whether you are ready to start, try the following test, which is based on the pre-knowledge items. You should not spend more than 20 minutes on the test. Hand your answers to your teacher for marking.

1. **a** Name two substances that make water 'hard'. (2)
   b Using the formula NaSp to represent a soap, write an equation for the reaction between soap and hard water. (2)
2. Write structural formulae for the following complex ions and state their shapes:
   a tetraamminecopper(II), (2)
   b tetrachlorocuprate(II). (2)
3. **a** Write a simple definition of a buffer solution in terms of its function. (2)
   b Name two types of substance (in addition to water!) found in most buffer solutions. (2)
4. Use the equilibrium law to derive the expression for the pH of a buffer solution:

$$pH = pK_a - \log \frac{[acid]}{[base]}$$ (4)

(Total: 16 marks)

CHAPTER 1

CARBOXYLIC ACIDS

■ 1.1 Names and structural formulae

Carboxylic acids contain the carboxyl group, $-CO_2H$ (or $-COOH$), which itself contains a **carb**onyl group and a hyd**roxyl** group; hence the name. The carboxyl group can be attached to an aliphatic or aromatic system as shown.

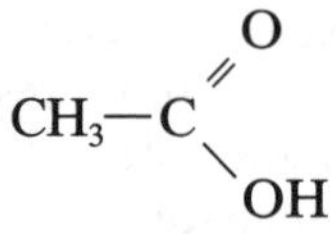

$$CH_3-C \overset{O}{\underset{OH}{\diagdown}}$$

ethanoic acid

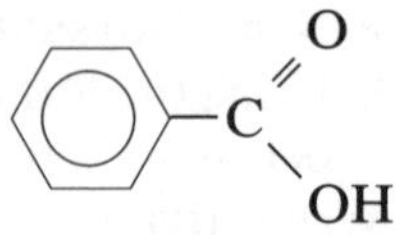

benzoic acid

OBJECTIVE

When you have finished this section you should be able to:
■ write the **structural formulae** and **names** of **carboxylic acids**.

Read the introductory chapter or section on carboxylic acids in your textbook(s) and look for the systematic names of carboxylic acids. You should then be able to do the next two exercises.

EXERCISE 1
Answers on page 163

Write the condensed structural formulae of:
a benzene-1,2-dicarboxylic acid,
b butanoic acid,
c 2-hydroxypropanoic acid (lactic acid),
d 2-hydroxybenzoic acid (salicylic acid),
e 3-methylpentanoic acid,
f ethanedioic acid (oxalic acid).

EXERCISE 2
Answers on page 163

Name the following compounds:
a $CH_3CH_2CO_2H$
b $C_6H_5CH_2CO_2H$
c $CH_3CH(CH_3)CH_2CO_2H$

d

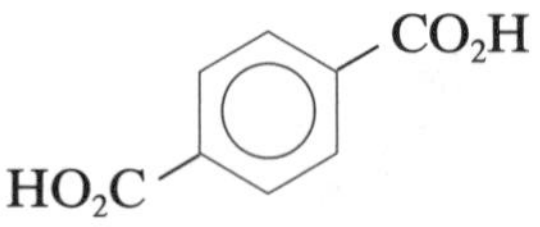

$$HO_2C-\bigcirc-CO_2H$$

e $CO_2H(CH_2)_3CO_2H$

As we have shown above, the carboxyl group consists of two reactive groups which you have already studied. In the next section, you find out the extent to which carboxylic acids behave like carbonyl compounds and like hydroxy compounds.

You will find it helpful to revise the reactions of carbonyl compounds and hydroxy compounds from ILPAC 8, Functional Groups, in order to make comparisons with carboxylic acids.

■ 1.2 Chemical properties of carboxylic acids

OBJECTIVES

When you have finished this section you should be able to:
■ describe some simple chemical tests which characterise carboxylic acids;
■ **compare** the reactions of carboxylic acids with **alcohols** and **carbonyl compounds**.

In the next exercise you make a prediction about the reactions you might expect of carboxylic acids based on your knowledge of alcohols and carbonyl compounds. Following that is an experiment to test your predictions.

EXERCISE 3

Answers on page 163

Predict what you might expect to observe when each of the following is added to a compound containing a C = O group and an OH group:
a sodium,
b sodium hydrogencarbonate solution,
c phosphorus pentachloride,
d 2,4-dinitrophenylhydrazine,
e iodine followed by sodium hydroxide solution,
f neutral iron(III) chloride solution,
g universal indicator solution.

In the next experiment, you test your predictions from Exercise 3.

EXPERIMENT 1

Chemical properties of carboxylic acids

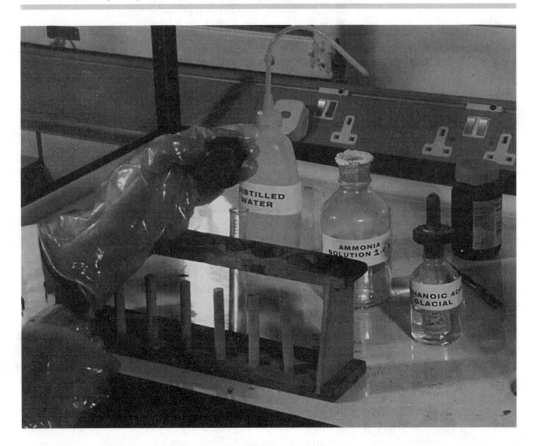

Aim

The purpose of this experiment is to see if carboxylic acids show the typical reactions both of alcohols and of carbonyl compounds.

Introduction

You will be carrying out some test-tube reactions on ethanoic acid. Ethanoic acid melts at 17 °C and therefore freezes in cold weather. Because solid ethanoic acid looks like ice, it is often described as 'glacial'. The reactions to be investigated are:
A. pH of aqueous solution.
B. Reaction with sodium hydrogencarbonate solution.
C. Reaction with sodium.
D. Reaction with phosphorus pentachloride.
E. Reaction with 2,4-dinitrophenylhydrazine.
F. Triiodomethane (iodoform) reaction.
G. Action of iron(III) chloride.

Requirements
- safety spectacles and gloves
- 5 test-tubes in rack
- wash-bottle of distilled water
- universal indicator solution
- ethanoic acid, glacial, CH_3CO_2H
- sodium hydrogencarbonate solution, saturated, $NaHCO_3$
- limewater, $Ca(OH)_2(aq)$
- forceps
- sodium, Na (1 mm cube under oil)
- filter papers
- wood splint
- phosphorus pentachloride, PCl_5
- spatula
- ammonia solution, 2 M NH_3
- 2,4-dinitrophenylhydrazine solution
- iodine solution, 10% I_2 in KI(aq)
- sodium hydroxide solution, 2 M NaOH
- sodium ethanoate (acetate), CH_3CO_2Na
- iron(III) chloride solution, 0.1 M $FeCl_3$
- Bunsen burner and bench mat

HAZARD WARNING

  Glacial ethanoic acid is flammable and corrosive.

 Sodium hydroxide solution is corrosive.

 Sodium and phosphorus pentachloride are extremely reactive, especially with water.

 2,4-Dinitrophenylhydrazine (Brady's reagent) is toxic.
Therefore you **must**:
- **wear safety spectacles and gloves;**
- **keep stoppers on bottles as much as possible;**
- **keep bottles of flammable liquids away from flames;**
- **keep sodium and phosphorus pentachloride away from water and moist air.**

**Procedure
– Part A**

pH of aqueous solution
1. Into a test-tube, pour about 2 cm^3 of distilled water and add one drop of universal indicator solution. Shake gently.
2. Add a few drops of glacial ethanoic acid, shake gently and note your observations in a copy of Results Table 1.

– Part B

Reaction with sodium hydrogencarbonate solution
Into a test-tube, add about 1 cm^3 of sodium hydrogencarbonate solution followed by a few drops of glacial ethanoic acid. Shake gently and test any gas produced.

– Part C

**Reaction with sodium
(Work at a fume cupboard, with your teacher present.)**
1. Pour about 2 cm^3 of glacial ethanoic acid into a **dry** test-tube in a rack standing in a fume cupboard.
2. Using forceps, pick up a 1 mm cube of sodium and blot it free of oil on some filter paper.

3. Drop the clean piece of sodium into the glacial ethanoic acid. Test any gas evolved and note your observations. Do not wash away the mixture until you are sure that the sodium has reacted completely. If some sodium remains unreacted, add a little more glacial ethanoic acid; on no account add water.

– Part D   Reaction with phosphorus pentachloride
1. Pour about 1 cm^3 of glacial ethanoic acid into a dry test-tube in a rack standing in a fume cupboard.
2. A little at a time, carefully add a spatula measure of phosphorus pentachloride. Bring the wet stopper of an ammonia bottle close to the mouth of the tube and note your observations.

– Part E   Reaction with 2,4-dinitrophenylhydrazine
Into a test-tube, pour about 2 cm^3 of 2,4-dinitrophenylhydrazine solution and about five drops of glacial ethanoic acid. Shake gently and note your observations.

– Part F   Triiodomethane (iodoform) reaction
Into a test-tube, place ten drops of iodine solution, and five drops of glacial ethanoic acid, followed by sodium hydroxide solution added dropwise until the colour of iodine disappears and a straw-coloured solution remains. Shake the tube gently and note your observations.

– Part G   Reaction with neutral iron(III) chloride solution
Into a test-tube, place about 2 cm^3 of sodium ethanoate solution (this avoids having to neutralise the acid) and a few drops of iron(III) chloride solution. Shake gently and then heat the solution. Note your observations.

Results Table 1
Reactions of ethanoic acid

Reaction	Observations
A. pH of aqueous solution	
B. Reaction with sodium hydrogen-carbonate solution	
C. Reaction with sodium	
D. Reaction with phosphorus pentachloride	
E. Reaction with 2,4-dinitro-phenylhydrazine	
F. Triiodomethane reaction	
G. Action of iron(III) chloride	

Specimen results on page 163

Questions
Answers on page 163

1. Were the predictions you made in Exercise 3 verified by experiment?
2. Does ethanoic acid more closely resemble hydroxy or carbonyl compounds in its reactions? Explain.
3. Which test indicates that ethanoic acid is a stronger acid than phenol?
4. Explain the fact that ethanoic acid is very soluble in water, whereas benzoic acid, $C_6H_5CO_2H$, a solid, is only slightly soluble in water.

The last experiment shows that, although carboxylic acids have some properties typical of hydroxy compounds, they have virtually nothing in common with aldehydes or ketones. We can deduce, therefore, that the close proximity of the hydroxyl and carbonyl groups causes a considerable interaction between the two. This is why the $—CO_2H$ group is treated as a separate functional group with its own characteristic reactions.

■ 1.3 Reactions of carboxylic acids

OBJECTIVE

When you have finished this section you should be able to:
■ write equations for the **reactions of carboxylic acids**, giving specific reaction conditions.

Read about the reactions of carboxylic acids in your textbook(s). Look for equations for the reactions you saw in the last experiment which gave positive results and for further reactions mentioned in the next exercise.

EXERCISE 4

Answers on page 164

Complete the following equations for the reactions of carboxylic acids, naming the products and stating necessary conditions.
a $CH_3CO_2H + NaOH(aq) \rightarrow$
b $CH_3CH_2CO_2H + Na_2CO_3 \rightarrow$
c $CH_3CO_2H + (NH_4)_2CO_3 \rightarrow$ (two stages) Check with your teacher, or syllabus, whether you need to know this particular reaction.
d $C_6H_5CO_2H + PCl_5 \rightarrow$
e $CH_3CH_2CO_2H \rightarrow$ (reduction)
f $CH_3CH_2CO_2H + CH_3OH \rightarrow$

The reactions illustrated in the last exercise are typical reactions of all carboxylic acids. The red colour which is produced when aliphatic (alkyl) carboxylic acids react with neutral iron(III) chloride, however, serves to distinguish them from aromatic acids, which give a buff precipitate.

The simplest carboxylic acid, methanoic acid, HCO_2H, shows some anomalous behaviour. Its structure suggests that it might have some of the properties of aldehydes as well as acids, and it does indeed reduce Tollens reagent. Note, however, that it does **not** undergo condensation reactions which are typical of aldehydes and ketones. Unless your syllabus states otherwise you will not be required to know the atypical properties of methanoic acid.

You now make a comparative summary of some of the reactions of typical carboxylic acids, hydroxy compounds and carbonyl compounds which give observable changes and can be used as distinguishing tests.

■ 1.4 Distinguishing tests

OBJECTIVE

When you have finished this section you should be able to:
■ describe simple **chemical tests** that you would use to **distinguish** between **carboxylic acids, hydroxy compounds** and **carbonyl compounds**.

EXERCISE 5

Answers on page 164

Complete a copy of Table 1, which compares the reactions of some carboxylic acids, hydroxy and carbonyl compounds.

Table 1

A comparative summary

Compound	Observation when tested with various reagents			
	Neutral iron(III) chloride	Sodium hydrogen-carbonate	2,4-dinitro-phenyl-hydrazine	Tollens reagent
Ethanoic acid CH_3CO_2H				
Benzoic acid $C_6H_5CO_2H$				
Ethanol C_2H_5OH				
Phenol C_6H_5OH				
Ethanal CH_3CHO				

In the two exercises that follow, you use these tests to distinguish between carboxylic acids, hydroxy and carbonyl compounds.

EXERCISE 6

Answers on page 165

Describe one simple reaction to distinguish between members of each of the following pairs of compounds.

a   CHO   and   CO$_2$H

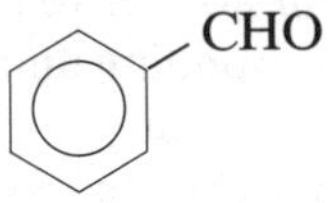

b   CO$_2$H   and   OH

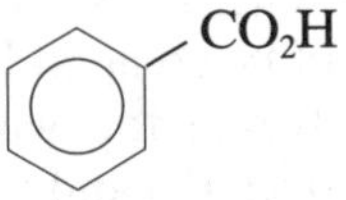

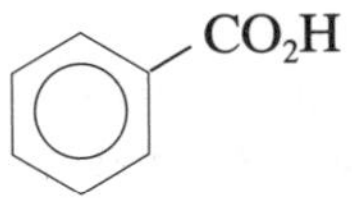

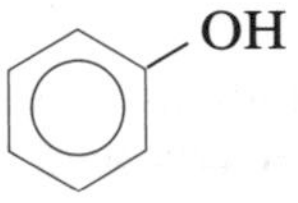

c   CH$_3$CO$_2$H   and   CO$_2$H

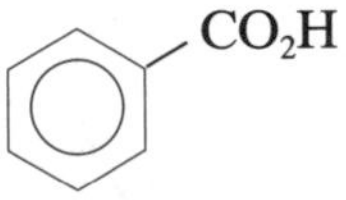

EXERCISE 7

Answers on page 165

The compounds ethanol, phenol, ethanoic acid and benzoic acid contain the hydroxyl group, — OH.

a   i)   Which two of these compounds react with sodium hydrogencarbonate?
    ii)   Write a balanced equation for the reaction involving one of them.
b   i)   Which one of the four compounds reacts with aqueous sodium hydroxide but does not react with sodium hydrogencarbonate?
    ii)   Write a balanced equation for this reaction.
c   i)   Which one of these compounds reacts with sodium metal but does not react with sodium hydrogencarbonate or with sodium hydroxide?
    ii)   Write a balanced equation for this reaction.

We now take a close look at the structure of the carboxyl group in order to explain the acidity of carboxylic acids.

■ 1.5 Acidity

OBJECTIVE When you have finished this section you should be able to:
■ account for the **acidity of carboxylic acids**.

The most important property of carboxylic acids is that they are weak acids. You will recall from ILPAC 8, Functional Groups, in the section on phenol that there are two principal factors that govern acidity. The acidity of a compound containing a hydroxyl group X—OH is increased firstly if X is electron-withdrawing and, secondly, if the anion XO⁻ is stabilised by distributing its charge over two or more atoms.

In the next exercise, you consider these factors along with some physical evidence about the structure of the carboxylate ion, RCO_2^-, in order to explain the acidity of carboxylic acids.

EXERCISE 8
Answers on page 165

The carboxylate ion is conventionally represented as:

$$CH_3C\underset{O^-}{\overset{O}{<}}$$

However, electron diffraction measurements have shown the ion to be symmetrical, and the distance of each oxygen atom from the carbon atom is intermediate between the normal lengths of C—O and C=O bonds.

a What does this evidence suggest about the conventional way of representing the carboxylate ion?

b Suggest a better representation of the carboxylate ion.

c Using an orbital diagram, show how your answer to (b) is possible. (Hint: see the orbital diagram of benzene in ILPAC 3, Bonding and Structure.)

d Considering the two factors which govern acidity, together with the diagrams of the carboxylate ion from (b) and (c), explain why ethanoic acid is a stronger acid than ethanol.

As well as delocalisation in the carboxylate ion, RCO_2^-, there must also be considerable delocalisation in carboxylic acid molecules, RCO_2H, as shown in Fig. 1.

Figure 1
Delocalisation in a carboxylic acid.

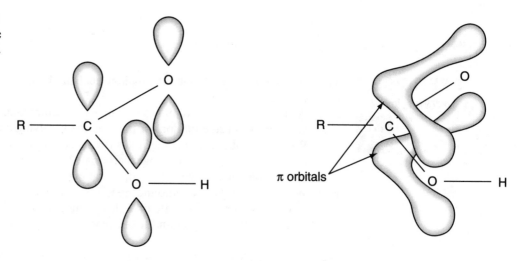

This delocalisation helps to explain why carboxylic acids undergo hardly any of the reactions we associate with the carbonyl group.

You have seen how the acidity of the hydroxyl group is very much enhanced by the carbonyl group next to it. An extension of this concept is that acid strength is affected by electron-attracting or electron-withdrawing substituents in the alkyl or aryl group next to the carboxyl group. This is illustrated in the next exercise.

EXERCISE 9

Answers on page 166

Table 2 lists the values of the dissociation constants of some carboxylic acids and substituted carboxylic acids. Study them and attempt the questions that follow.

Table 2

Name	Formula	K_a/mol dm^{-3}
Benzoic acid	$C_6H_5CO_2H$	6.3×10^{-5}
Methanoic acid	HCO_2H	1.6×10^{-4}
Ethanoic acid	CH_3CO_2H	1.7×10^{-5}
Propanoic acid	$CH_3CH_2CO_2H$	1.3×10^{-5}
Chloroethanoic acid	CH_2ClCO_2H	1.3×10^{-3}
Dichloroethanoic acid	$CHCl_2CO_2H$	5.0×10^{-2}
Trichloroethanoic acid	CCl_3CO_2H	2.3×10^{-1}
2-Chlorobutanoic acid	$CH_3CH_2CHClCO_2H$	1.4×10^{-3}
3-Chlorobutanoic acid	$CH_3CHClCH_2CO_2H$	8.7×10^{-5}
4-Chlorobutanoic acid	$CH_2ClCH_2CH_2CO_2H$	3.0×10^{-5}

a Explain the trend in the acid strength of ethanoic, chloroethanoic, dichloroethanoic and trichloroethanoic acids.

b Arrange the following acids in order of decreasing acid strength, and explain the reasons for your suggested order.

$$ClCH_2CO_2H \quad BrCH_2CO_2H \quad ICH_2CO_2H \quad FCH_2CO_2H \quad CH_3CO_2H$$

c Explain the trend in the acid strength of 2-chlorobutanoic, 3-chlorobutanoic and 4-chlorobutanoic acids.

d Explain the trend in the acid strength of methanoic, ethanoic and propanoic acids.

e Explain why benzoic acid is a stronger acid than ethanoic acid.

We now consider the various methods of preparing carboxylic acids.

■ 1.6 Methods of preparation

OBJECTIVE

When you have finished this section you should be able to:
■ write equations for **reactions in which carboxylic acids are produced**, giving specific reaction conditions.

Read about the methods of preparing carboxylic acids and hydroxy-carboxylic acids. Also, look for the most recent industrial preparations of methanoic, ethanoic, benzoic and ethanedioic (oxalic) acids – these have some commercial uses. This will enable you to do the next exercise.

EXERCISE 10
Answers on page 167

Complete the equations in a copy of Table 3, which summarises the reactions in which carboxylic acids are produced. Name the reactants and products and state the necessary conditions.

Table 3 General methods of preparing carboxylic acids

A. **Oxidation**
 1. Primary alcohols
    $CH_3CH_2OH + H_2O \rightarrow$

 2. Aldehydes
    $CH_3CHO + H_2O \rightarrow$ } write electronic half-equations

 3. Alkyl benzenes
    $C_6H_5CH_3 + 2H_2O \rightarrow$

B. **Hydrolysis**
 1. Nitriles
    $CH_3CH_2CN + 2H_2O + H^+ \rightarrow$

 2. Amides
    $CH_3CONH_2 + H_2O + H^+ \rightarrow$

 3. Acyl halides
    $CH_3COCl + H_2O \rightarrow$

 4. Anhydrides
    $(CH_3CO)_2O + H_2O \rightarrow$

 5. Esters
    $CH_3CH_2CO_2CH_3 + H_2O \rightarrow$

Preparation of hydroxycarboxylic acids

Cyanohydrin reaction
$CH_3CHO + HCN \rightarrow$

    $+ 2H_2O + H^+ \rightarrow$ } 2 stages

$CH_3COCH_3 + HCN \rightarrow$

    $+ 2H_2O + H^+ \rightarrow$ } 2 stages

Industrial methods

A. **Methanoic acid**
   $NaOH(aq) + CO(g) \rightarrow$

B. **Ethanoic acid**
   alkanes $+ O_2 \rightarrow$
   (mixtures from petroleum)

C. **Benzoic acid**
   $C_6H_5CH_3 + 1\frac{1}{2}O_2 \rightarrow$

D. **Ethanedioic acid**
   $2HCO_2Na \rightarrow$
       $+ 2H^+ \rightarrow$ } 2 stages

The earliest method for producing ethanoic acid involved the oxidation of ethanol by a bacterial process. In recent years, ethanoic acid has been produced by two different continuous processing methods. One of these uses the oxidation of hydrocarbons, whereas the other involves the reaction between methanol and carbon monoxide, in the presence of a catalyst. You compare these methods in the next teacher-marked exercise. Alternatively you could consider the answers to the questions as a class discussion with your teacher. You may wish to refresh your memory on the factors that must be considered when setting up a chemical process from ILPAC 8, Functional Groups, in Chapter 2 Industrial processes, page 37.

EXERCISE
Teacher-marked

Until recently, most UK ethanoic acid was produced by the naphtha oxidation method. Naphtha is one of the fractions obtained in petroleum refining. It is principally a mixture of alkanes, cycloalkanes and aromatic hydrocarbons which contain between five and nine carbon atoms per molecule. Naphtha is oxidised by air at a temperature of about 450 K and a pressure of around 50 atm. No catalyst is needed and the reaction takes place in one step. The product contains about 50% ethanoic acid together with a range of other compounds including propanone, methanoic acid, propanoic acid and butane-1,4-dioic acid. Some of these co-products are useful; others are simply burned at the chemical plant to provide energy.

Naphtha oxidation is gradually being replaced by a process in which methanol reacts with carbon monoxide in the presence of a rhodium/iodine catalyst.

$$CH_3OH + CO \rightarrow CH_3COOH \qquad \Delta H^\ominus = -135 \text{ kJ mol}^{-1}$$

This process takes place at a temperature similar to the one used in naphtha oxidation, but the pressure is lower (about 30 atm). The catalyst is expensive and costs about £25 g^{-1}, but the process is very specific. A 99% yield of ethanoic acid is obtained. Methanol is cheap and readily available from a number of sources. It can be made from coal, petroleum or natural gas.

a One way of investigating the success and efficiency of a chemical process is to look at the advantages and disadvantages associated with factors such as:
  ■ the conditions employed,
  ■ the feedstock,
  ■ the product,
  ■ the nature of the co-products.

For example, you can use this approach to explain why the methanol/carbon monoxide process is replacing naphtha oxidation.

Taking the four factors in turn, suggest **one** advantage in each case which the methanol/carbon monoxide process has over naphtha oxidation.

b Suggest one disadvantage which the methanol/carbon monoxide process has compared with naphtha oxidation.

You now summarise reactions and preparations of carboxylic acids.

■ 1.7 Summary of reactions and preparations

EXERCISE 11

Answer on page 168

Complete a copy of Fig. 2 to summarise the preparations and reactions of carboxylic acids.

Figure 2
Summary of preparations and properties of ethanoic acid.

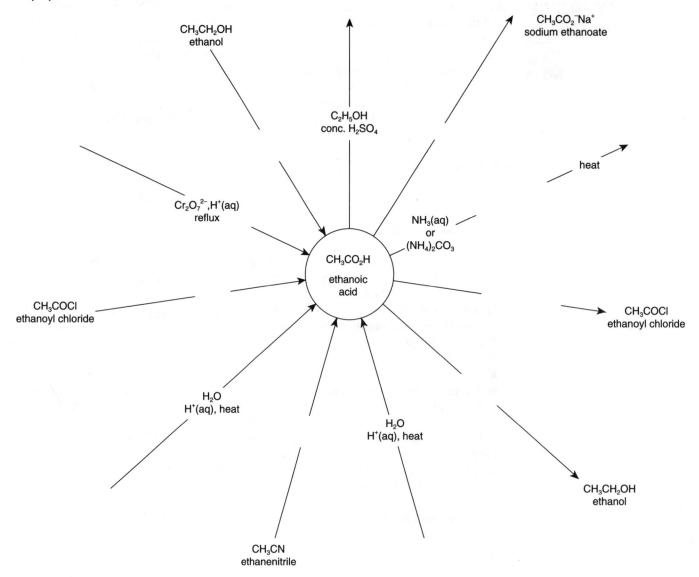

In the next section you devise some synthetic pathways which include the reactions and preparations of carboxylic acids; they also include reactions of other functional groups you have studied.

■ 1.8 Synthetic pathways

EXERCISE 12

Answers on page 168

The following sequence is proposed for the industrial preparation of a local anaesthetic (**F**), called **benzocaine**.

CH_3 → Step 1 → CH_3 / NO_2 (**A**) → Step 2 → CH_2Cl / NO_2 (**B**) → Step 3 → CH_2OH / NO_2 (**C**) → Step 4 →

$COOH$ / NO_2 (**D**) → Step 5 → $COOCH_2CH_3$ / NO_2 (**E**) → Step 6 → $COOCH_2CH_3$ / NH_2 (**F**)

Give the reagents, essential reaction conditions, and reaction type for **each** of the steps 1–6.

Products containing benzocaine.

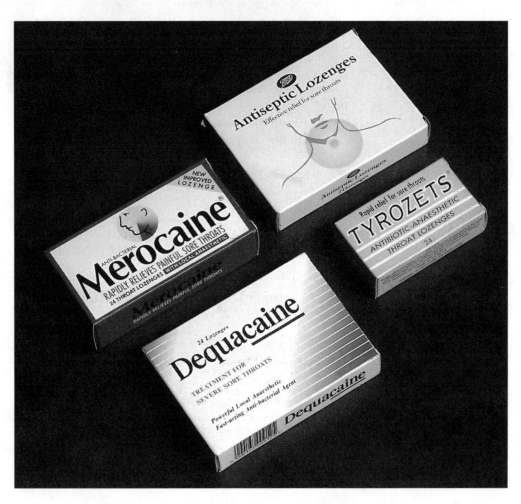

EXERCISE 13

Answers on page 168

Show how you would bring about the following conversions in no more than three stages.

a $C_6H_5NH_2 \rightarrow C_6H_5CO_2H$
b $CH_3CHO \rightarrow CH_3CH(OH)CO_2H$
c $CH_3(CH_2)_3I \rightarrow CH_3(CH_2)_3CO_2H$
d $CH_3(CH_2)_3CO_2H \rightarrow CH_3(CH_2)_4I$

EXERCISE 14

Answer on page 169

Jasmine flowers.

Phenylmethyl ethanoate is a constituent of jasmine flower oil and it is often used for soap aromas. Suggest a one-step synthesis for this ester:

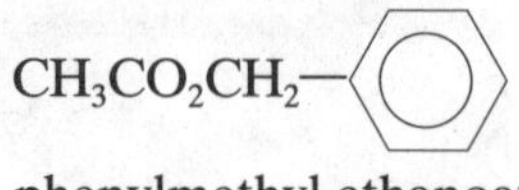

$$CH_3CO_2CH_2-\bigcirc$$

phenylmethyl ethanoate

We now consider some of the uses of carboxylic acids.

■ 1.9 Uses of carboxylic acids

OBJECTIVE

When you have finished this section you should be able to:
■ describe some of the uses of **carboxylic acids**.

Read about the uses of carboxylic acids in your textbook. In particular, look for applications in the manufacture of synthetic flavourings, nylon, food preservatives, vinegar and Terylene. We suggest that you do not learn the details of the procedures, but gain sufficient background knowledge to do the next exercise. Further study of nylon and Terylene is included later in this book.

EXERCISE 15
Answers on page 169

Ethanoic acid, hexanedioic (adipic) acid, benzoic acid, and benzene-1,4-dicarboxylic (terephthalic) acid are used in the manufacture of the useful materials shown in the photographs below. Match each carboxylic acid with one or more photographs.

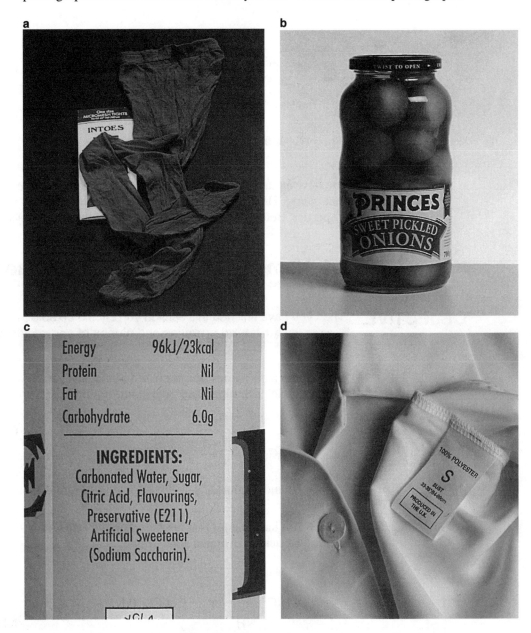

To consolidate your knowledge of the chemistry of carboxylic acids you should attempt the following teacher-marked exercise.

EXERCISE
Teacher-marked

In the light of your knowledge of the chemical reactions of carboxylic acids, discuss the truth of the following statement: 'The reactions of the carboxyl group are typical of those of the carbonyl and hydroxyl groups that it contains'.

Illustrate your answer with suitable reactions.

You have now completed a study of carboxylic acids. We suggest that you spend some time revising what you have learned, particularly Fig. 2 on page 14, before you go on.

In the next chapter we deal with acyl halides, amides, acyl anhydrides and esters, all of which are grouped together as carboxylic acid derivatives.

CHAPTER 2

CARBOXYLIC ACID DERIVATIVES

In this chapter, you study carboxylic acid derivatives. You may regard them as compounds in which the hydroxyl group of a carboxylic acid has been replaced by another functional group. Four different classes of compounds that fall into this category are illustrated below.

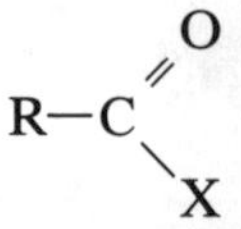

$$R-C\underset{X}{\overset{O}{\diagup}} \qquad R-C\underset{NH_2}{\overset{O}{\diagup}} \qquad R-C\underset{OR'}{\overset{O}{\diagup}} \qquad R-C\overset{O}{\diagup}$$

an acyl halide 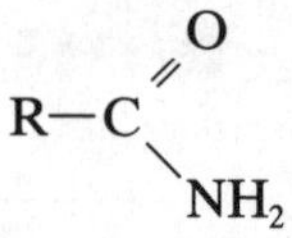   an amide 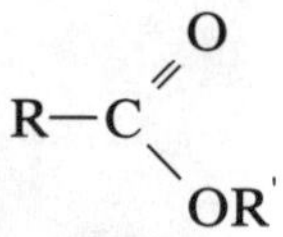   an ester

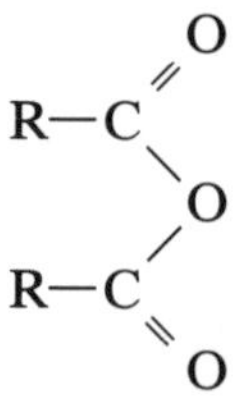

$$R-C\overset{O}{\diagdown}$$

an acyl anhydride

(X = a halogen, R = alkyl or aryl group)

You have already met many examples of carboxylic acid derivatives in various parts of your course. Here we consider them together because they have some common properties. Before we look at these we clarify the naming system.

■ 2.1 Names and structural formulae

When you have finished this section you should be able to:
■ write the **structural formulae** and **names of carboxylic acid derivatives**.

Acyl halides have names that are based on the acids from which they are derived. The suffix 'oic acid' is dropped and '-oyl halide' is added.

e.g. $CH_3C\underset{Cl}{\overset{O}{\diagup}}$ or CH_3COCl $\bigcirc\!\!\!\!-C\underset{Cl}{\overset{O}{\diagup}}$ or C_6H_5COCl

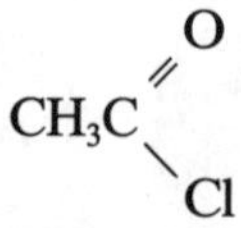

       ethanoyl chloride   benzoyl chloride

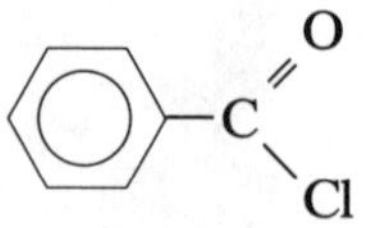

Amides also have names that are based on the acids from which they are derived. The ending '-oic acid' is dropped and '-amide' added.

e.g. $CH_3C\underset{NH_2}{\overset{O}{\diagup}}$ or CH_3CONH_2 $\bigcirc\!\!\!\!-C\underset{NH_2}{\overset{O}{\diagup}}$ or $C_6H_5CONH_2$

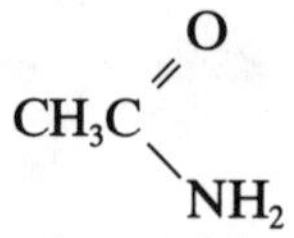

       ethanamide   benzamide

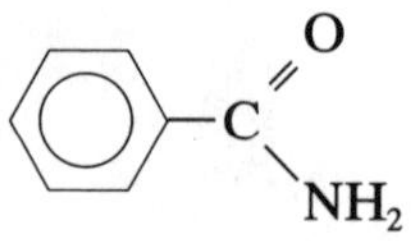

Acyl anhydrides are named as derivatives of the parent acid (or acids) by replacing the word 'acid' with 'anhydride'.

e.g. $CH_3C\overset{O}{\diagup}$
           O or $(CH_3CO)_2O$ $C_6H_5C\overset{O}{\diagup}$
      $CH_3C\overset{}{\diagdown}$ O or $C_6H_5CO_2COCH_3$
          O $CH_3C\overset{}{\diagdown}$
                                                   O

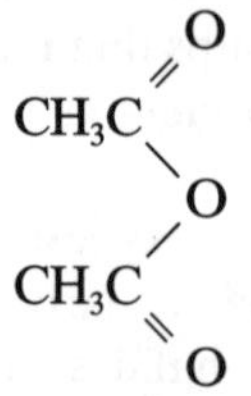

      ethanoic anhydride   benzoic ethanoic anhydride

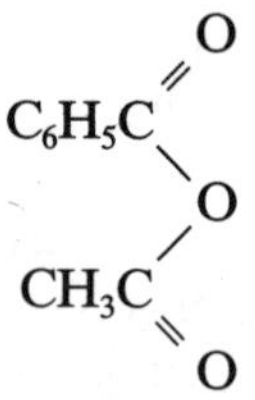

Esters are named as alkyl or aryl derivatives of the parent carboxylic acid. The ending '-ic' is dropped and '-ate' is added.

e.g.

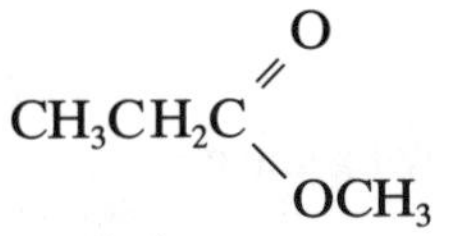

$$CH_3CH_2C \overset{O}{\underset{OCH_3}{\big<}}$$

or $CH_3CH_2CO_2CH_3$ or $CH_3OCOCH_2CH_3$

methyl propanoate

To help you name the esters correctly, we give you a worked example.

WORKED EXAMPLE   Name the esters $CH_3CO_2CH_2CH_2CH_3$ and $CH_3OCOCH_2CH_2CH_3$.

Solution

1. Identify the $C = O$ group and the two atoms (one C, one O) to which it is attached.
2. Draw a line through the formula to separate the CO_2 group, with its attached C atom, from the rest of the molecule.

$$CH_3CO_2 | CH_2CH_2CH_3 \quad CH_3 | OCOCH_2CH_2CH_3$$

If you are in doubt, draw the structural formula.

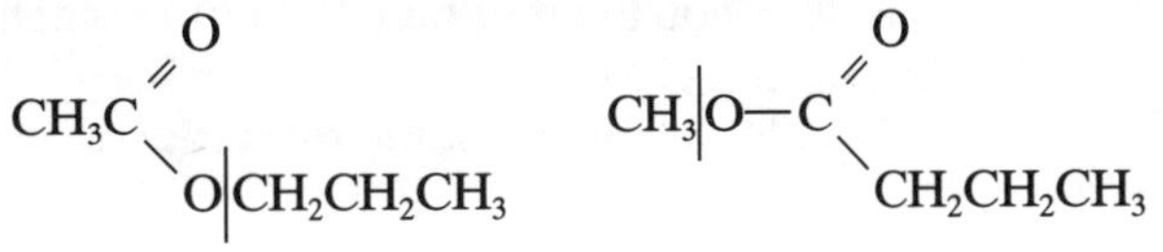

$$CH_3C \overset{O}{\underset{O|CH_2CH_2CH_3}{\big<}} \qquad CH_3|O - C \overset{O}{\underset{CH_2CH_2CH_3}{\big<}}$$

3. Name the portion without the CO_2 group as an alkyl radical.

$$CH_3CO_2 | \underset{\text{propyl}}{CH_2CH_2CH_3} \quad \underset{\text{methyl}}{CH_3} | OCOCH_2CH_2CH_3$$

4. Name the portion with the CO_2 group as a carboxylate radical.

$$\underset{\text{ethanoate}}{CH_3CO_2} | CH_2CH_2CH_3 \quad CH_3 | \underset{\text{butanoate}}{OCOCH_2CH_2CH_3}$$

5. Combine the two names from steps 3 and 4 with the alkyl part of the name first.

$$\underset{\text{propyl ethanoate}}{CH_3CO_2 | CH_2CH_2CH_3} \quad \underset{\text{methyl butanoate}}{CH_3 | OCOCH_2CH_2CH_3}$$

You can reverse the process to write the formula from a given name:

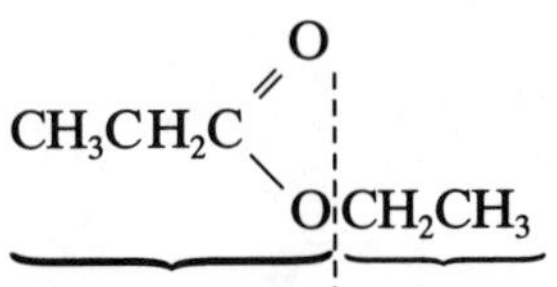

$$CH_3CH_2C \overset{O}{\underset{\underset{\text{propanoate} \,|\, \text{ethyl}}{O|CH_2CH_3}}{\big<}}$$

ethyl propanoate

In the next exercise, you write names and formulae for some esters and other derivatives of carboxylic acids.

EXERCISE 16
Answers on page 169

a Name the following substances:
i) $HCO_2CH_2CH_3$
ii) C_2H_5COCl
iii) $C_2H_5CONH_2$

iv) $CH_3COCCH_2CH_3$
      $\underset{O}{\overset{\|}{}}$ $\underset{O}{\overset{\|}{}}$

v) $C_6H_5CO_2CH_2CH_3$

b Write formulae for the following substances:
i)   propyl ethanoate,
ii) butanamide,

iii) benzoic ethanoic anhydride,
iv) phenyl benzoate.

Carboxylic acid derivatives have many reactions in common, as you see in the next section.

■ 2.2 Chemical properties of carboxylic acid derivatives

OBJECTIVES

When you have finished this section you should be able to:
■ write equations for the **reactions of carboxylic acid derivatives** with **nucleophilic reagents**;
■ describe the **mechanism** for the reaction between carboxylic acid derivatives and nucleophiles;
■ write equations for the **reduction** of carboxylic acid derivatives.

A feature common to all carboxylic acid derivatives is the attachment of an electron-withdrawing group, Y, next to the carbon atom.

$$R-\overset{\delta^+}{\underset{\overset{\|}{O^{\delta^-}}}{C}}-Y \qquad (Y\ is\ -Hal, -OCOR', -OR'\ or\ NH_2)$$

This makes the derivatives more susceptible than their parent carboxylic acid to nucleophilic attack at the carbonyl carbon atom.

We now consider the reactions between carboxylic acid derivatives and nucleophilic reagents.

■ 2.3 Nucleophilic substitution of acid derivatives

We can represent the reactions between carboxylic acid derivatives and nucleophilic reagents by the general equation shown in Fig. 3, where HZ is the nucleophilic reagent.

Figure 3

$$R-\underset{\overset{\|}{O}}{C}-Y \ + \ HZ \longrightarrow R-\underset{\overset{\|}{O}}{C}-Z \ + \ HY$$

In the next experiment you illustrate the reactions of carboxylic acid derivatives, taking ethanoyl chloride as an example.

EXPERIMENT 2   Chemical properties of ethanoyl chloride

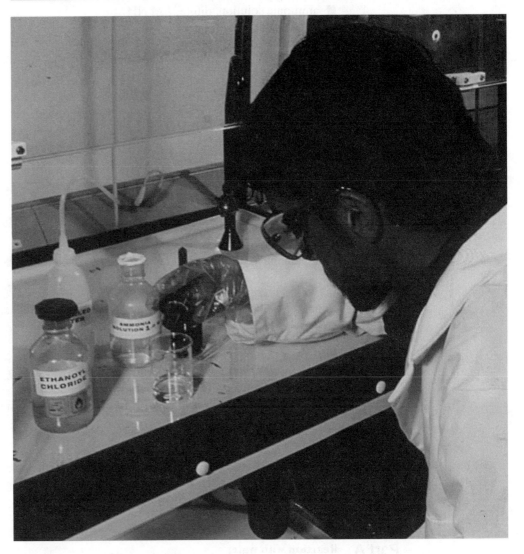

Aim   The purpose of this experiment is to illustrate the chemistry of carboxylic acid derivatives by investigating the reactions of ethanoyl chloride with some nucleophilic reagents.

Introduction   You or your teacher will be carrying out the following reactions of ethanoyl chloride – a typical acyl halide:

A. Reaction with water.
B. Reaction with ethanol.
C. Reaction with ammonia.
D. Reaction with phenylamine.

Since some of the reactions are violent, your teacher may decide to demonstrate this experiment. If you do it yourself, you must take great **care**!

Requirements
■ safety spectacles
■ protective gloves
■ 4 beakers, 100 cm^3
■ wash-bottle of distilled water

- 4 teat-pipettes
- ethanoyl chloride, CH₃COCl
- ammonia solution, dilute, 2 M NH₃
- glass rod
- universal indicator paper
- iron(III) chloride solution, 0.1 M FeCl₃
- test-tube
- sodium carbonate solution, 1 M Na₂CO₃
- ethanol, C₂H₅OH
- ammonia solution, concentrated, '0.880' NH₃
- phenylamine, C₆H₅NH₂

HAZARD WARNING

 Ethanoyl chloride gives off a vapour which burns the skin, eyes and respiratory tract. Some of its reactions are violent.

 Phenylamine is very poisonous if inhaled, swallowed or absorbed through the skin.

 Concentrated aqueous ammonia burns the skin and the vapour irritates the eyes.

 Ethanol is flammable. Therefore you **must**:
- **wear safety spectacles and gloves;**
- **keep stoppers on the bottles as much as possible;**
- **perform the experiment in a fume cupboard protected by a safety glass.**

Procedures   The reactions with ammonia and phenylamine are particularly violent. **You must follow the instructions carefully.** Look again at the hazard warning and only perform the tests in a fume cupboard protected by a safety glass.

– Part A   **Reaction with water**
1. Pour about 5 cm³ of distilled water into a small beaker placed in the fume cupboard with the front pulled down as far as is practicable.
2. Wearing safety spectacles, add a few drops of ethanoyl chloride to the water. Bring the moist stopper of a bottle of ammonia (2M NH₃, **not** the concentrated 0.880 NH₃) solution near to the top of the beaker and note your observations.
3. Neutralise the solution in the beaker by adding dilute aqueous ammonia (2M NH₃, **not** the concentrated 0.880 NH₃) dropwise until a drop of the solution from a glass rod gives a neutral colour to universal indicator.
4. In a test-tube, neutralise about 1 cm³ of iron(III) chloride solution by adding sodium carbonate solution dropwise until a faint precipitate just remains on shaking.
5. Add a few drops of the neutral iron(III) chloride to the neutral solution prepared in step 3 and note your observations in a copy of Results Table 2.

– Part B   **Reaction with ethanol**
1. Repeat steps 1 and 2 of part A using ethanol instead of water in the beaker.
2. Add sodium carbonate solution until there is no further effervescence and note the smell of the product.

– Part C **Reaction with ammonia**
Repeat steps 1 and 2 of part A using concentrated aqueous ammonia instead of water in the beaker. Take special care when you add the ethanoyl chloride a drop at a time. Note your observations.

– Part D **Reaction with phenylamine**
1. Pour five drops of phenylamine in a small beaker placed in the fume cupboard with the front pulled down as far as is practicable.
2. Wearing safety spectacles, add a few drops of ethanoyl chloride to the phenylamine. Note your observations.

Results Table 2

Reaction	Observations
A. **Reaction with water** Product tested with ammonia Product tested with iron(III) chloride	
B. **Reaction with ethanol** Product tested with ammonia Smell of product	
C. **Reaction with ammonia**	
D. **Reaction with phenylamine**	

Specimen results on page 169

Questions
Answers on page 169

1. Using the general equation for the nucleophilic substitution reactions of carboxylic acid derivatives in Fig. 3, page 20, together with your observations from the experiment, write equations for the reactions of ethanoyl chloride and the following nucleophilic reagents. Name the products.
   a Water, H_2O,
   b Ethanol, C_2H_5OH,
   c Ammonia, NH_3,
   d Phenylamine, $C_6H_5NH_2$.
2. Explain the order of reactivity of the four nucleophiles towards ethanoyl chloride. This will be easier to answer after the next section.

In the next exercise you write similar equations for the other carboxylic acid derivatives.

EXERCISE 17
Answers on page 170

Using Fig. 3 as a guide, complete the equations in a copy of Table 4 which lists some reactions of carboxylic acid derivatives with nucleophilic reagents. Name the reactants and products and state essential conditions.

Table 4 Reactions of carboxylic acid derivatives with nucleophiles	A. **With water**		Conditions
	1. Acyl halides	e.g. $C_2H_5COCl + H_2O \rightarrow$	
	2. Acyl anhydrides	e.g. $(CH_3CO)_2O + H_2O \rightarrow$	
	3. Esters	e.g. $C_2H_5CO_2CH_3 + H_2O \rightarrow$	
	4. Amides	e.g. $CH_3CONH_2 + H_2O \rightarrow$	
	B. **With alcohols and phenols**		
	1. Acyl halides	e.g. $C_2H_5COCl + C_2H_5OH \rightarrow$	
	2. Acyl anhydrides	e.g. $CH_3CO_2COCH_3 + C_6H_5OH \rightarrow$	
	No reaction with esters or amides		
	C. **With ammonia**		
	1. Acyl halides	e.g. $C_2H_5COCl + 2NH_3 \rightarrow$	
	2. Acyl anhydrides	e.g. $C_2H_5CO_2COC_2H_5 + NH_3 \rightarrow$	
	3. Esters	e.g. $CH_3CO_2C_2H_5 + NH_3 \rightarrow$	
	No reaction with amides		
	D. **With amines**		
	1. Acyl halides	e.g. $C_2H_5COCl + CH_3NH_2 \rightarrow$	
	2. Acyl anhydrides	e.g. $(CH_3CO)_2O + C_3H_7NH_2 \rightarrow$	
	3. Esters	e.g. $CH_3CO_2C_2H_5 + C_2H_5NH_2 \rightarrow$	
	No reaction with amides		

Note that the hydrolysis of an ester or an amide by aqueous alkali yields a salt rather than the carboxylic acid obtained by acid hydrolysis. However, these salts are readily converted to the carboxylic acid by the addition of mineral acid.

$$CH_3CO_2C_2H_5 + OH^- \rightarrow CH_3CO_2^- + C_2H_5OH$$

$$CH_3CONH_2 + OH^- \rightarrow CH_3CO_2^- + NH_3$$

The fact that amides produce ammonia on warming with alkali serves to distinguish them from amines.

Judging by the number of reactions listed in Table 4, it would appear that acyl halides and acyl anhydrides are more reactive than esters, with amides being the least reactive towards nucleophilic reagents.

In practice, it is found that acyl halides react far more vigorously than acyl anhydrides, so the order of reactivity of the carboxylic acid derivatives towards nucleophilic reagents is:

acyl halide > acyl anhydride > ester > amide

This corresponds to the known electron-attracting power of the substituent groups:

$$-Hal > -CO_2R > -OR > -NH_2$$

Many of the reactions of carboxylic acid derivatives are important in synthesis. As you can see, the reactions of acyl halides and acyl anhydrides are similar in that they both replace a hydrogen atom with an acyl group, RCO—, and consequently both are described as acylating agents. Since acyl anhydrides are not so violently reactive as acyl halides, they are sometimes preferred to them for making amides and esters. One important application of this is in the manufacture of 2-ethanoyloxybenzenecarboxylic acid (aspirin).

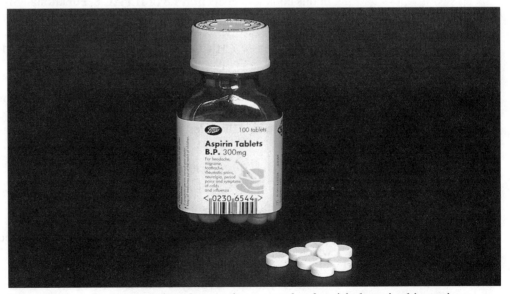

You will have the opportunity of preparing a sample of aspirin later in this section.

We now consider the mechanism for the reaction between carboxylic acid derivatives and nucleophilic reagents. Your particular syllabus may not require you to know the details of the mechanisms of these reactions. Your teacher will be able to tell you whether you can skip this section.

■ 2.4 Mechanism for nucleophilic substitution of carboxylic acid derivatives

Before attempting the next exercise you should refresh your memory of the mechanism of addition–elimination reactions of carbonyl compounds (discussed in ILPAC 8, Functional Groups, and the mechanism of nucleophilic substitution reactions of alkyl halides (Section 7.4, ILPAC 5, Introduction to Organic Chemistry).

The two-stage reaction sequence shown below shows the reaction between a nucleophile, HZ:, and a general carboxylic acid derivative, CH_3COY. Study it and attempt the exercise that follows.

Figure 4

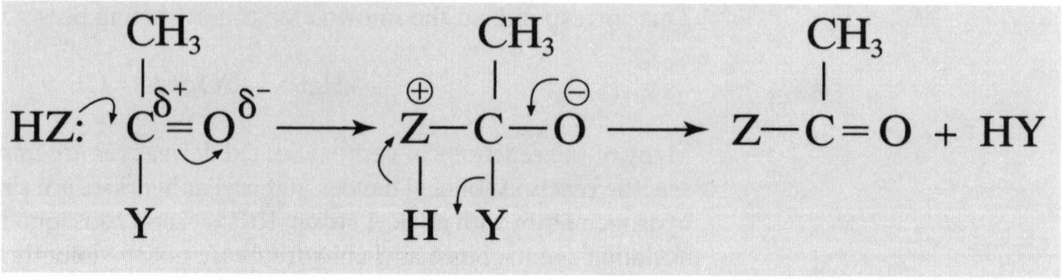

$$HZ: \quad \overset{CH_3}{\underset{Y}{\overset{|}{C}}} \overset{\delta+}{=} \overset{\delta-}{O} \longrightarrow \overset{CH_3}{\underset{\underset{H \quad Y}{|}}{\overset{\oplus}{Z}-\overset{|}{C}-\overset{\ominus}{O}}} \longrightarrow \overset{CH_3}{\underset{}{Z-\overset{|}{C}=O}} + HY$$

EXERCISE 18
Answers on page 171

a Using Fig. 4 as a guide, write a sequence for the reaction between ethanoyl chloride and water.
b Classify each stage of the sequence you have just written.
c How does the mechanism of these reactions compare with the mechanisms for:
  i) addition–elimination (condensation) reactions of carbonyl compounds,
  ii) substitution reactions in alkyl halides?

Another example of this mechanism is the hydrolysis of esters, which is the reverse of the esterification reaction you studied as one of the reactions of carboxylic acids earlier in this book and in Section 1.8 of ILPAC 8, Functional Groups.

$$CH_3CO_2C_2H_5 + H_2O \underset{\text{esterification}}{\overset{\text{hydrolysis}}{\rightleftharpoons}} CH_3CO_2H + C_2H_5OH$$

You may wish to refresh your memory on this before attempting the next exercise.

EXERCISE 19
Answer on page 172

On hydrolysis of the ester, a bond could be broken at either (1) or (2) in the following diagram:

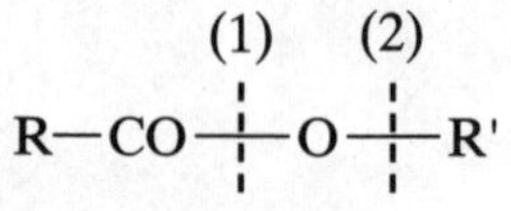

$$\overset{(1) \quad (2)}{R-CO \vdash O \vdash R'}$$

Suggest a method by which, using an isotope of oxygen, you could identify the point at which bond-breaking occurs.

We have seen how carboxylic acid derivatives show similarities in their reactions with nucleophilic reagents. Another similarity is in their reactions with reducing agents.

■ 2.5 Reduction of carboxylic acid derivatives

As with the parent carboxylic acids, reduction of the derivatives can be brought about with suitable reducing agents. (See ILPAC 8, Functional Groups, Table 10, page 30 on choice of reducing agents.)

Read about the reactions of acyl halides, amides, esters and acyl anhydrides with reducing agents in your textbook. You should then be able to do the next exercise.

EXERCISE 20
Answers on page 172

Complete the equations in a copy of Table 5. Name the reagents that can perform all the reductions.

1.	Acyl halides	e.g. $CH_3COCl + H_2 \rightarrow$
2.	Acyl anhydrides	e.g. $(CH_3CO)_2O + 4H_2 \rightarrow$
3.	Esters	e.g. $CH_3CH_2CO_2CH_3 + 2H_2 \rightarrow$
4.	Amides	e.g. $CH_3CONH_2 + 2H_2 \rightarrow$

Amides undergo two other reactions that are important in organic synthesis. These are the Hofmann degradation reaction and dehydration. If these reactions are not a requirement of your particular syllabus, skip them and proceed to Section 2.7.

■ 2.6 Other reactions of amides

OBJECTIVE

When you have finished this section you should be able to:
■ write equations for the **Hofmann degradation** reaction of amides and the **dehydration of amides**.

Read about the reaction of amides with bromine in the presence of alkali (Hofmann degradation) and find out why this reaction is of particular importance in organic synthesis. Also find out the conditions that are necessary for dehydrating amides to nitriles. You should then be able to do Exercises 21 and 22.

EXERCISE 21
Answers on page 172

a Complete the following equation for a Hofmann degradation reaction:

$$CH_3CH_2CONH_2 + Br_2 + 4OH^- \rightarrow$$

b Why is this reaction of particular importance in organic synthesis?

You will use this reaction in synthetic pathways later in this chapter.

EXERCISE 22
Answer on page 172

Write an equation to show how ethanamide may be dehydrated to ethanenitrile. State the necessary conditions.

You have already studied the reactions of alkyl halides in ILPAC 5, Introduction to Organic Chemistry, in the section covering halogen-compounds. By comparing their reactions with those of acyl halides you will see how the presence of a carbonyl group has a profound effect on the properties of the latter.

■ 2.7 Comparing acyl halides and alkyl halides

OBJECTIVE   When you have finished this section you should be able to:
■ **compare the reactivity of acyl halides and alkyl halides** and relate the differences to their structure.

Before you attempt the next exercise, refresh your memory on the reactions of alkyl halides from ILPAC 5, Introduction to Organic Chemistry (Section 7.9), and the reactions of acyl halides from Experiment 2 earlier in this book. Note the ways in which primary alkyl halides and acyl halides differ in their reactions with nucleophilic reagents such as water and ammonia.

EXERCISE 23
Answers on page 172

a Does ethanoyl chloride, CH_3COCl, react more or less vigorously than chloroethane, C_2H_5Cl, in its reactions with the nucleophilic reagents water and ammonia?
b How does the charge on the carbon atom in the $—COCl$ group, compare with the charge on the carbon atom in the $—CCl$ group?
c Use your answers from (a) and (b) to explain the difference in reactivity between acyl halides and alkyl halides.

We now consider the effect of the carbonyl group on the reactivity of the $—NH_2$ group. We do this by comparing the properties of amides and amines.

■ 2.8 Comparing amides and amines

OBJECTIVES   When you have finished this section you should be able to:
■ describe how the presence of the carbonyl group modifies the properties of the adjacent $—NH_2$ group.
■ explain why **amides are weaker bases** than amines.

If you compare the reactions of amides with the reactions of amines you studied in ILPAC 8, Functional Groups, you will see that they have very little in common. The amino group of the amide must therefore undergo considerable interaction with the carbonyl group. This is further illustrated by the fact that amides are much weaker bases than amines, being neutral to litmus and only forming unstable salts with strong acids.

In the next section, you consider how the presence of the carbonyl group in amides has the effect of reducing their basic strength relative to amines.

■ 2.9 Basicity

You will recall from your study of amines that their basicity is determined by the availability of the lone pair of electrons on the nitrogen atom and by the relative stability of the amine, RNH_2, and its conjugate acid, RNH_3^+. Basic properties are lessened if the amine is relatively more stable than its conjugate acid.

In the next exercise you consider these factors, together with an orbital diagram of the $—CONH_2$ group, in order to explain why amides are weaker bases than amines.

EXERCISE 24

Answer on page 173

It is believed that, in amides, the lone pair of electrons on the nitrogen atom interacts with the π orbitals of the carbonyl bond, giving an extended delocalised system.

Fig. 5 shows how the individual p orbitals interact to give the extended delocalised system.

Figure 5
Delocalisation in an amide.

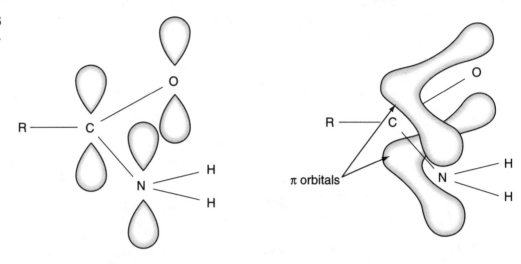

In what way does this orbital diagram help to explain why amides are weaker bases than amines?

Comparing the base strength of aliphatic amines, aromatic amines and amides is a popular examination question – as the following exercise illustrates. The question includes some revision of your earlier work on amines in ILPAC 8, Functional Groups.

EXERCISE 25

Answers on page 173

a Arrange the following in order of increasing base strength (i.e. the least basic first): ammonia, phenylamine (aniline), butylamine and ethanamide.

b Explain the difference in basicity between:
  i)   butylamine and phenylamine,
  ii) butylamine and ethanamide,
  iii) ethanamide and phenylamine.

Distinguishing tests, as you know, often appear in A-level questions. You now describe some simple chemical tests that would enable you to distinguish between carboxylic acid derivatives and compounds containing similar groups.

■ 2.10 Distinguishing tests

OBJECTIVE   When you have finished this section you should be able to:
■ describe some **simple chemical tests** to **distinguish** between **carboxylic acid derivatives** and compounds containing similar groups.

Use your textbook, if necessary, to find out how the reaction with sodium hydroxide serves to distinguish between amides and amines. Also find out how the reaction with water serves to distinguish between acyl halides and alkyl halides, and how different esters may be distinguished by means of hydrolysis. This should enable you to do the next exercise.

EXERCISE 26
Answers on page 174

Describe how you would distinguish by chemical means between:

a chlorobenzene, C_6H_5Cl; 1-chlorobutane, C_4H_9Cl; and ethanoyl chloride, CH_3COCl.

b

$$\text{C}_6\text{H}_5\!-\!\overset{\displaystyle O}{\underset{\displaystyle OC_2H_5}{C}} \quad \text{and} \quad CH_3\!-\!\overset{\displaystyle O}{\underset{\displaystyle OC_2H_5}{C}}$$

c

$$\text{C}_6\text{H}_5\!-\!\overset{\displaystyle O}{\underset{\displaystyle OC_2H_5}{C}} \quad \text{and} \quad \text{C}_6\text{H}_5\!-\!\overset{\displaystyle O}{\underset{\displaystyle O\!-\!C_6H_5}{C}}$$

We now consider the various methods of preparing carboxylic acid derivatives. You have met most of these reactions already as reactions of other functional groups.

■ 2.11 Methods of preparation

OBJECTIVE

When you have finished this section you should be able to:

■ write equations for the **preparation of carboxylic acid derivatives**, giving specific reaction conditions.

Look up methods of preparing acyl halides, amides, esters and acyl anhydrides in your textbook. This will enable you to do the next exercise.

EXERCISE 27
Answers on page 175

a Complete the equations in a larger copy of Table 6, which summarises the reactions in which carboxylic acid derivatives are produced. Name the reactants and products and state any special reaction conditions.

Table 6
Formation of derivatives of carboxylic acids

A.	**Acyl halides**	
1.	From carboxylic acids	e.g. $CH_3CO_2H + ? \rightarrow$
		(3 different reagents)

B.	**Amides**	
1.	From ammonium salts	e.g. $CH_3CH_2CO_2^- NH_4^+ \rightarrow$
2.	From acyl halides	e.g. $C_3H_7COCl + ? \rightarrow$
3.	From esters	e.g. $CH_3CO_2C_2H_5 + ? \rightarrow$
4.	From anhydrides	e.g. $(CH_3CO)_2O + ? \rightarrow$

C.	**Esters**	
1.	From carboxylic acids and hydroxy compounds	e.g. $CH_3CO_2H + C_2H_5OH \rightarrow$
2.	From acyl halides and hydroxy compounds	e.g. $C_2H_5COCl + C_6H_5OH \rightarrow$
3.	From anhydrides and hydroxy compounds	e.g. $(CH_3CO)_2O + C_2H_5OH \rightarrow$

b In the preparation of acyl halides from carboxylic acids, why is sulphur dichloride oxide (thionyl chloride) often used in preference to other reagents? (Hint: look at the nature of the inorganic products.)

In the next experiment you prepare the ester 2-ethanoyloxybenzenecarboxylic acid, more commonly known as aspirin. Since your teacher may use this for practical assessment we do not give specimen results. In Appendix 1 we have included a similar experiment to prepare the ester phenyl benzoate, from phenol and benzoyl chloride. Ask your teacher which of these experiments you should do.

 The ILPAC video programme 'Organic Techniques II', shows you the processes of recrystallisation, filtration and melting point determination. If you are not already familiar with these techniques you should either view this video tape or ask your teacher for a demonstration.

EXPERIMENT 3   Preparation of aspirin

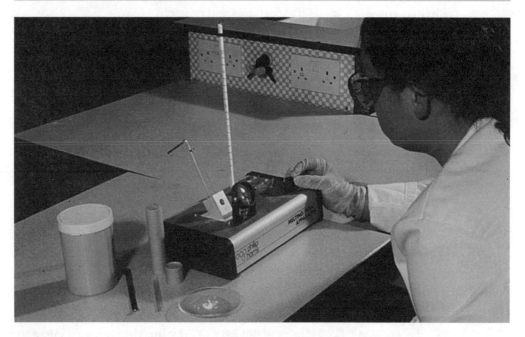

Aim   The purpose of this experiment is to prepare a sample of aspirin (2-ethanoyloxy-benzenecarboxylic acid), purify it by recrystallisation, measure its melting point and estimate the yield. You may wish to refer back to ILPAC 5, Introduction to Organic Chemistry, page 95 for the method of calculating percentage yield.

Introduction   You prepare aspirin by heating together 2-hydroxybenzoic acid and ethanoic anhydride in the presence of phosphoric(V) acid (catalyst) on a steam bath:

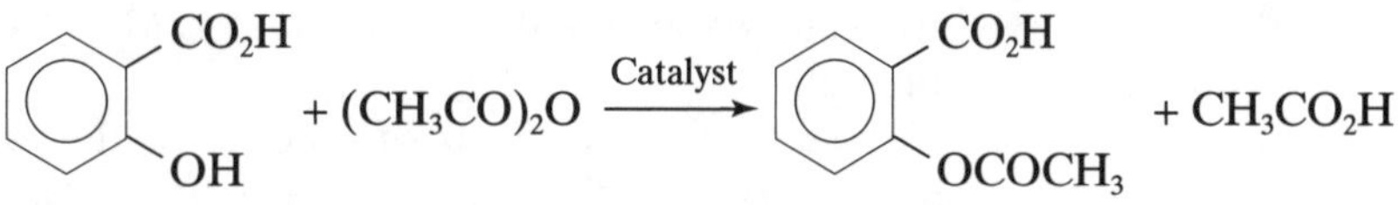

$$\text{2-hydroxybenzoic acid} + (CH_3CO)_2O \xrightarrow{\text{Catalyst}} \text{2-ethanoyloxybenzenecarboxylic acid (aspirin)} + CH_3CO_2H$$

2-ethanoyloxybenzene-
carboxylic acid
(aspirin)

The product appears as a solid which you purify by filtering and recrystallising from water.

HAZARD WARNING

  Ethanoic anhydride is corrosive and flammable. It is particularly dangerous to eyes.

 2-hydroxybenzoic acid is harmful to eyes, lungs and skin.

 Phosphoric(V) acid (orthophosphoric acid) is corrosive and must be kept in a tray to avoid spillage.
Therefore you **must:**
- **wear safety glasses and protective gloves.**
Any aspirin that is prepared must **not** be used medicinally.

Requirements – Part A
- safety spectacles and protective gloves
- weighing bottle
- spatula
- ethanoic anhydride $(CH_3CO)_2O$, 4 cm^3
- 2-hydroxybenzoic acid (salicylic acid), 11 g
- phosphoric(V) acid, 85%, a few drops (with teat-pipette)
- access to balance, sensitivity 0.01 g
- measuring cylinder, 5 cm^3
- beaker, 250 cm^3
- beaker, 100 cm^3
- reflux apparatus consisting of Liebig condenser and 50 cm^3 pear-shaped flask (Fig. 6(a))
- conical flask, 100 cm^3
- glass rod
- suction filtration apparatus (see Fig. 6(b))
- ice, about half a 250 cm^3 beaker of crushed ice
- wash-bottle of distilled water

Procedure – Part A

Preparation of aspirin
1. Transfer about 2.0 g of 2-hydroxybenzoic acid into a weighing bottle and weigh to the nearest 0.01 g.
2. Into a 50 cm^3 pear-shaped flask pour 4 cm^3 of ethanoic anhydride (**care**) and the bulk of the 2-hydroxybenzoic acid from the weighing bottle.
3. Reweigh the weighing bottle, with any remaining solid, to the nearest 0.01 g.
4. To the mixture in the pear-shaped flask add five drops of 85% phosphoric(V) acid (**care – return to spillage tray with used teat-pipette**) and swirl to mix.
5. Fit the flask with a reflux condenser as shown in Fig. 6(a) (remember there is no stopper in the top of the condenser) and heat the mixture on the steam bath (beaker of boiling water) for about five minutes. Make sure the solid is carefully swirled at intervals.
6. Without cooling, add 2 cm^3 of water down the condenser. This will hydrolyse any excess ethanoic anhydride and is a vigorous reaction.
7. When the reaction from step 6 has finished pour the mixture into 40 cm^3 of cold water in a 100 cm^3 beaker and allow to cool to room temperature.
8. Complete crystallisation by placing the small beaker and contents into an ice bath. If necessary, scratch the sides of the beaker with a glass rod to induce crystallisation.
9. Collect the product by suction filtration (Fig. 6(b)) and wash with a little distilled water.

Figure 6

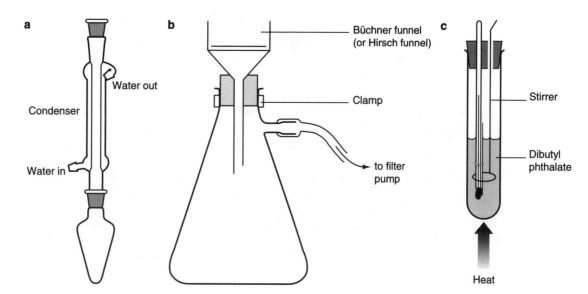

a Condenser / Water out / Water in

b Büchner funnel (or Hirsch funnel) / Clamp / to filter pump

c Stirrer / Dibutyl phthalate / Heat

Requirements – Part B

- boiling-tube
- glass rod
- water-bath or 250 cm³ beaker
- Bunsen burner, tripod, gauze and bench mat
- thermometer, 0–250 °C
- ice
- suction filtration apparatus (see Fig. 6(b))
- filter papers
- specimen bottle
- access to balance, sensitivity 0.01 g

Procedure – Part B

Recrystallisation

1. Transfer the crystals to a boiling-tube and just cover them with distilled water.
2. Place the boiling-tube in a beaker of hot water, kept at about 60 °C, and stir with a glass rod.
3. If some solid is still visible, add just enough water to dissolve it completely after stirring.
4. Cool the solution in an ice–water mixture until crystals appear.
5. Filter the crystals through the suction apparatus, using a clean Büchner funnel and filter paper. To avoid losing any solid, break the vacuum and use the filtrate to rinse the boiling-tube into the funnel.
6. Using suction again, rinse the crystals with about 1 cm³ of cold distilled water and drain thoroughly.
7. Press the crystals between two wads of filter paper to remove excess solvent. Then put the crystals on another dry piece of filter paper placed alongside (not on top of) a Bunsen burner and gauze, turning the crystals over occasionally until they appear dry. Don't let them get too hot or they will melt!
8. Weigh the dry crystals in a pre-weighed specimen bottle and record the mass of your sample of aspirin. Calculate the percentage yield using the method described in ILPAC 5, Introduction to Organic Chemistry, page 95, Section 7.7, and the expression:

$$\% \ yield = \frac{actual \ mass \ of \ product}{maximum \ mass \ of \ product} \times 100$$

Requirements – Part C

- melting-point tubes (at least 2)
- watch-glass
- thermometer, 0–250 °C, long stem

COLEG MENAI
BANGOR, GWYNEDD LL57 2TP

- boiling-tube fitted with cork and stirrer (see Fig. 6(c)) ⎫ or electrical
- dibutylbenzene-1,2-dicarboxylate (dibutyl phthalate) ⎪ melting
- retort stand, boss and clamp ⎬ point
- rubber ring or band ⎭ apparatus

**Procedure
– Part C**

Determination of melting point

1. Take a melting-point tube and push the open end through a pile of aspirin on a watch glass, until a few crystals have entered. If the crystals are large, you may need to crush or grind them first.
2. Tap the closed end of the tube vertically against a hard surface, or rub with the milled edge of a coin, to make the solid fall to the bottom.
3. Repeat the filling and tapping procedure until a total length of 0.2–0.5 cm is compacted at the bottom of the tube. Prepare another tube in this way. If you have an electrical melting point apparatus go to step 7. If not go to step 4.
4. Attach one of the prepared melting-point tubes to the thermometer, as shown in Fig. 6(c).
5. Half-fill the boiling-tube with dibutyl phthalate and position the thermometer with attached tube and the stirrer through the bung, as shown in Fig. 6(c).
6. Position the apparatus over a micro-burner (or low Bunsen flame) and gauze and **gently** heat the apparatus, stirring the dibutyl phthalate all the time by moving the stirrer up and down.
7. Keep an eye on the crystals and note the temperature as soon as signs of melting are seen (usually seen as a contraction of the solid followed by a damp appearance). Record the range of temperature over which your sample melts. This first reading gives only a rough melting point but is a guide for the second determination.
8. Remove the burner and the old tube containing aspirin. Allow the temperature to drop about 10 °C before positioning a fresh melting-point tube containing another portion of the aspirin.
9. Repeat the above procedure in order to obtain a more accurate value of the melting point. Raise the temperature very slowly (about 2 °C rise per minute) until the crystals melt (take the formation of a visible meniscus as a sign of melting).

Hand in your product suitably labelled, i.e. mass, % yield, melting point, name and date.

Results Table 3

Mass of weighing bottle + 2-hydroxybenzoic acid	g
Mass of weighing bottle after emptying	g
Mass of 2-hydroxybenzoic acid	g
Mass of specimen bottle	g
Mass of specimen bottle + aspirin	g
Mass of aspirin	g
Melting point of aspirin	°C

The compounds that you used to prepare aspirin in the last experiment are the same as those used in the actual industrial method of production. In the exercise that follows you consider the costs involved in producing aspirin.

EXERCISE
Teacher-marked

a Using a chemicals catalogue work out the cost of the materials used in producing a 300 mg tablet of aspirin by the method used in Experiment 3. Compare this with the current price of a 300 mg tablet bought at a chemist. With the rest of your class and teacher you should discuss the reasons for any price differences. Your discussion should consider:
- problems of scale-up,
- cost of bulk chemicals compared with the small quantities used in your experiment,
- cost of setting up and maintaining a chemicals plant,
- cost of distribution and advertising.

b Although aspirin is a useful drug because it has analgesic (pain-killing) anti-inflammatory and antipyretic (fever-reducing) actions it also causes bleeding of the lining of the stomach. For this reason soluble aspirin is safer to use. What further reactions can you suggest would convert your sample of aspirin into soluble aspirin?

In the next section you consider the history behind the development of aspirin and its safe use.

■ 2.12 Aspirin – its development and safety

OBJECTIVE

When you have finished this section you should be able to:
■ evaluate information by extraction from text about the value to society of organic materials as medicines using aspirin as an example.

Aspirin was the first drug to be produced synthetically. Find out more about the development, safety and uses of this simple drug. Your teacher should be able to suggest some suitable references.

A typical aspirin tablet contains 0.3 g (300 mg) of aspirin. As with all medicines it is important to know the recommended doses and possible hazards. We have already discussed some of the hazards associated with taking aspirin in this and previous volumes. Now would be a good opportunity of bringing them together along with the benefits in a class discussion. Your discussion could extend to other medication as well. The following facts could help you plan your discussion.

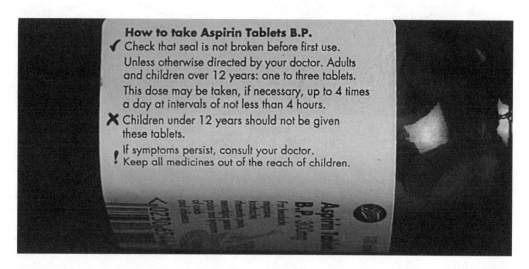

How to take Aspirin Tablets B.P.
✓ Check that seal is not broken before first use.
  Unless otherwise directed by your doctor. Adults and children over 12 years: one to three tablets.
  This dose may be taken, if necessary, up to 4 times a day at intervals of not less than 4 hours.
✗ Children under 12 years should not be given these tablets.
! If symptoms persist, consult your doctor.
! Keep all medicines out of the reach of children.

■ In the UK, over 200 people a year die of aspirin poisoning with children under five forming a large proportion of those who die by accidental poisoning.
■ Aspirin can cause internal bleeding. Up to 6 cm^3 of blood a day can be lost.
■ Clinical trials have shown that aspirin may be effective in treating heart attack and stroke victims.

Aspirin has been around for a long time. Bringing a new drug on to the market is a gamble – it takes on average 12 years of research and development and an investment of £125 million. Fewer than 5 out of 10 000 potential medicines ever reach a hospital pharmacy or chemist's shop. Figure 7 shows a breakdown of the many costs that go into producing a new drug which are then passed on to the customer.

Figure 7
Breakdown of the costs
involved in producing a new
drug.

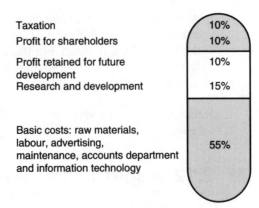

You now summarise the reactions and preparations of carboxylic acid derivatives. Because of the very close relationship between the derivatives of a parent carboxylic acid, the reactions are best shown on a single chart.

■ 2.13 Summary of reactions and preparations

EXERCISE 28

Answers on page 176

Complete a copy of Fig. 8, which summarises the reactions and preparations of carboxylic acid derivatives.

Figure 8
Summary of
interconversions of carboxylic
acid derivatives.
For other reactions
see Figure 2.

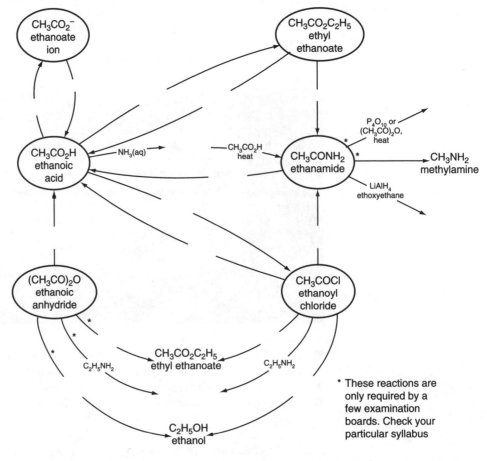

In the next section you write some synthetic pathways which include the reactions and preparations of carboxylic acids and carboxylic acid derivatives. Two of the pathways in Exercise 30, (a) and (f), involve the removal of a carbon atom from a compound: attempt these only if the Hofmann degradation reaction appears in your syllabus.

■ 2.14 Synthetic pathways

EXERCISE 29

Answer on page 176

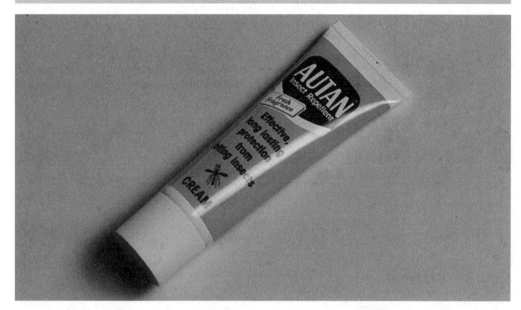

The ester phenylmethyl benzoate $\langle \rangle - \overset{\overset{\displaystyle O}{\displaystyle \|}}{C} - O - CH_2 - \langle \rangle$ is present in

some insect repellents. Starting from methylbenzene only, suggest a synthesis for this ester.

EXERCISE 30

Answers on page 176

Suggest synthetic pathways suitable for bringing about the following conversions. State the reaction conditions for each step and indicate the intermediate products.
(Only do (a) and (f) if your syllabus specifies the Hofmann degradation reaction.)

a $C_3H_7CO_2H \rightarrow C_3H_7NH_2$
b $CH_3CHO \rightarrow CH_3CONH_2$
c $CH_3CONH_2 \rightarrow CH_3CO_2CH_2CH_3$
d $C_6H_5CO_2H \rightarrow C_6H_5CO_2C_2H_5$
e $CH_3CO_2H \rightarrow C_2H_5CO_2H$
f $C_3H_7OH \rightarrow C_2H_5OH$

EXERCISE 31

Answers on page 177

Devise a reaction scheme for the preparation of *N*-phenylbenzamide, $C_6H_5CONHC_6H_5$, from benzene and methylbenzene. No other organic compound is available but you may use whatever inorganic reagents you wish.

For each step in the synthesis you should give the names of the reagents, the conditions of reaction and the structures of intermediate compounds. Balanced equations will not be expected.

EXERCISE 32

Answers on page 177

This question is about organic reactions.

a The scheme below shows two routes by which ethanol may be converted into ethylamine.

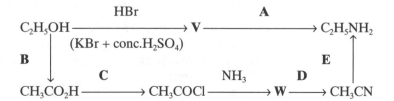

i) Identify organic substances **V** and **W** by writing their structural formulae.
ii) Write down the reagents used in stages **A–E**.
iii) Identify the type of reaction that occurs in stage **B**.

b What is the formula of the organic compound formed when $C_2H_5NH_2$ reacts with CH_3COCl?

EXERCISE 33

Answers on page 177

Benzoic acid, $C_6H_5CO_2H$, is used as a preservative in some fruit products. It is harmless to humans in small quantities because it can be excreted in the urine as hippuric acid, **D**:

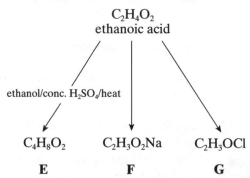

D

a Name two functional groups in **D**.
b Give the formulae of two compounds which you could react together in the laboratory to make **D**.
c What reagents and conditions would be needed to make a sample of benzoic acid from **D**?

EXERCISE 34

Answers on page 178

The reaction scheme below gives the molecular formulae of compounds which can be obtained from ethanoic acid, $C_2H_4O_2$.

$$C_2H_4O_2$$
ethanoic acid

ethanol/conc. H_2SO_4/heat

$C_4H_8O_2$	$C_2H_3O_2Na$	C_2H_3OCl
E	**F**	**G**

a Give the name and the structural formula of **each** of the compounds, **E**, **F** and **G**.
b State the reagent(s) and reaction conditions which could be used for converting ethanoic acid into:
i) **F**,
ii) **G**.
c Ethanoic acid can be obtained from **E**.
i) State the reaction conditions required.
ii) Write a balanced chemical equation for the reaction.
iii) Give the name of the type of reaction which occurs.
iv) What process may be used to separate the products?
d i) Write an equation for the reaction between **G** and phenylamine, $C_6H_5NH_2$.
ii) To what class of compounds does the organic product belong?

Carboxylic acid derivatives have a variety of uses which you may care to read about. Esters, for instance, are widely used as solvents for paints and varnishes and synthetic flavourings. Acyl anhydrides such as *cis*-butenedioic (maleic) anhydride, when reacted with poly-functional alcohols such as ethan-1,2-diol, form polyesters. You study polyesters as well as polyamides later in this book in the chapter covering synthetic polymers.

You have now completed a study of the major functional groups in organic chemistry. You will meet some of them again in later sections of this book when you study amino acids, proteins and synthetic polymers. Now would be a good time to bring together and revise all the functional groups you have met earlier in this book as well as in ILPAC 5, Introduction to Organic Chemistry, and ILPAC 8, Functional Groups.

■ 2.15 A suggestion for revision

One way of achieving this is by constructing a large summary chart of all the organic reactions you should know. You do this in the next teacher-marked exercise.

EXERCISES
Teacher-marked

1. Using a large piece of paper, at least A3 size, construct a summary chart which incorporates and links together the chemical reactions of all the functional groups you have studied in this and previous organic books. You will need to consult your completed summary charts of the various functional groups you have studied and link them together in one single chart. It is also very useful if you identify the type of mechanism by symbol, colour code or otherwise, for all those reactions whose mechanisms are specified on your particular syllabus. You may find it helpful to put this chart on your bedroom wall for frequent reference.

You can now test your knowledge of organic reactions by attempting the next question. Since you are required to suggest tests for some given compounds you will find Table 21 Characteristic tests for functional groups on page 111 of ILPAC 8, Functional Groups, useful. This will also help you prepare for the experiment that follows.

2. You are supplied with unlabelled samples of six organic compounds. You are also provided with the following reagents:
   1. 2,4-Dinitrophenylhydrazine solution.
   2. Ammoniacal silver nitrate solution.
   3. Red litmus paper.
   4. Dilute sodium hydroxide solution.
   5. Dilute hydrochloric acid solution.

Complete a copy of Table 7, indicating initially the results you would expect to **observe** on interacting each of the organic compounds separately with reagents 1 and 2. Then devise a scheme of three further tests on each compound using **only** the other reagents listed above, so that you would be able to identify all six organic compounds. Any test which you suggest **must** lead to some **observable** result. In the spaces marked *, indicate the reagent(s) and conditions used.

Table 7

Unlabelled compound		Reagent 1	Reagent 2	*	*	*
Name	Formula					
Propanal	C_2H_5CHO					
Propanone	CH_3COCH_3					
Methyl benzoate	$C_6H_5COOCH_3$					
Ethanamide	CH_3CONH_2					
Ethanoic acid	CH_3COOH					
1-Aminopropane	$CH_3CH_2CH_2NH_2$					

We complete this section by asking you to identify, by experiment, some organic substances. This is both useful revision of organic reactions and can either be used as preparation for a practical examination or as a practical assessment.

If it is available, view the second part of the ILPAC video programme 'Identifying unknown substances' before you do the practical test. This will help you not only with the practical technique but also with the interpretation of observations.

If you are allowed to refer to notes and textbooks during the practical (ask your teacher) you may find it useful to refer to Table 21, Characteristic tests of functional groups, on page 111 of ILPAC 8, Functional Groups.

EXPERIMENT 4 Identification of two organic compounds

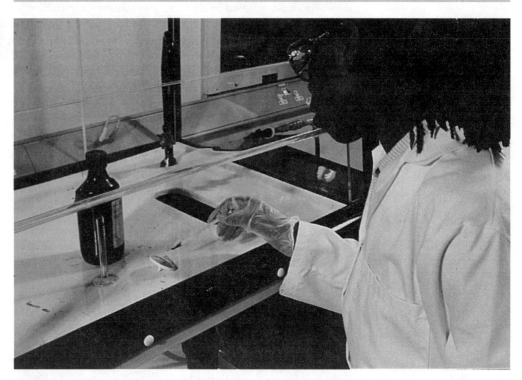

Aim The purpose of this experiment is to identify the functional groups present in two unknown organic compounds from observations made and inferences drawn from some simple chemical tests.

Introduction The procedure that follows is taken from two A-level practical examination papers. Since you should preferably work under examination conditions, all the apparatus and chemicals should be provided for you so we have not included a requirements list. However, take note of the hazard warnings.

HAZARD WARNING

  Concentrated hydrochloric acid and solutions of silver nitrate, sodium chlorate(I) (sodium hypochlorite) and sodium hydroxide are all corrosive.

  2,4-Dinitrophenylhydrazine is toxic by skin absorption. Therefore you **must**:
■ **wear safety spectacles and gloves throughout.**

  Compound G produces a harmful vapour. Therefore you **must**:
■ **perform experiments with these compounds at a fume cupboard;**
■ **keep tops on bottles as much as possible.**

  Compound G is flammable. Therefore you **must**:
■ **keep this compound away from flames.**

You are provided with two organic compounds, labelled F and G. Each compound contains the elements carbon, hydrogen and oxygen only. You are provided with an aqueous solution of F and a pure sample of G. Carry out the following experiments. Record your observations and inferences in (a larger copy of) the results table provided. Comment on the types of chemical reaction occurring and, where possible, deduce the functional groups present in these compounds. Then answer the questions that follow. Note that the procedure for the triiodomethane reaction in test (d) is different to the one we have followed in previous experiments, but it gives the same result.

Results Table 4

Experiment	Observations	Inferences
a To 2 cm^3 of the solution of F add 2,4-dinitrophenylhydrazine reagent		
b Prepare a sample of Tollens reagent as follows: to 5 cm^3 of aqueous silver nitrate in a test-tube add one to two drops of aqueous sodium hydroxide. Then add dilute aqueous ammonia until only a trace of precipitate remains. Now add five drops of the solution of F and place the tube in hot water. (Pour the contents of the tube down the sink on completion of the test)		
c Add a little of the solution of F to some sodium hydrogencarbonate		
d To 1 cm^3 of the solution of F add 3 cm^3 of aqueous potassium iodide and then 10 cm^3 of sodium chlorate(I) (sodium hypochlorite) solution		
e Place **one** drop of G on an inverted crucible lid and ignite G from above		
f To 1–2 cm^3 of G add 2,4-dinitrophenylhydrazine reagent		
g Prepare a sample of Tollens reagent as in (b). Add five drops of G, shake the mixture and place the tube in hot water. (Pour the contents of the tube down the sink on completion of the test)		

Questions
1. Comment on the structural features of F.
2. Comment on the structural features of G.

We now look at naturally occurring esters called lipids which play a major role in living systems.

LIPIDS

The term 'lipid' is applied to a wide range of substances that are insoluble in water but soluble in non-polar organic solvents. Most of these are esters that have large molecules, the most familiar being glycerides, often known simply as 'fats'.

OBJECTIVES   When you have finished this chapter you should be able to:
■ recognise the **main groups of lipid** and their **biological functions**;
■ interpret the **molecular structures of fats**;
■ explain the physical differences between **saturated** and **unsaturated fats** in terms of the structure of their carboxylic acids;
■ describe how **hydrogenation** of vegetable oil converts it into margarine;
■ explain the meaning of the term **iodine number** and how it can be determined.

There are four main groups of lipid.
1. **Glycerides** (fats). These are esters of glycerol (propane-1,2,3-triol) and long-chain carboxylic acids ('fatty acids'). Glycerides form the main energy storage material in plant and animal cells.
2. **Waxes.** These are mostly esters of long-chain monohydric alcohols and fatty acids. Waxes form protective coatings on skin, fur, feathers, leaves and fruits.
3. **Compound lipids such as phospholipids.** These can be regarded as glycerides with one of the ester groups replaced by another group, usually a nitrogenous base attached via a phosphate link. Phospholipids are important structural components of cell membranes.
4. **Steroids.** These are derived from the saturated hydrocarbon represented by the formula shown below.

They include the male and female sex hormones, the digestive bile acids and cholesterol. An excess of cholesterol has been linked with 'hardening of arteries' (arteriosclerosis) and heart disease.

We now focus on fats. The steroid cholesterol, a lipid of current consumer interest, is discussed in a later section that deals with health issues.

■ 3.1 Structure of fats and oils

 Fats and oils have the same basic structure. The only difference is that solids are called fats, while liquids are usually referred to as oils. Consult a textbook to find out the general structure of a molecule of a fat (also known as triglyceride, triacylglycerol or triester) and what happens when it is hydrolysed. You should also find out what part fats play in our diet. A biology textbook may be more useful for this.

 Fats and oils are esters of propane-1,2,3-triol (glycerol) with long-chain carboxylic acids (fatty acids). The following scheme (Fig. 9) shows how a fat or oil molecule could be formed from the esterification of propane-1,2,3-triol and three long-chain carboxylic acids. Study it and then attempt the exercise that follows. You will also need a model-making kit.

Figure 9

esterification → + 3H₂O

propane-1,2,3-triol (glycerol)

Three 'fatty acids' where R, R' and R" are long-chain hydrocarbons

general fat molecule (triester)

EXERCISE 35

Answers on page 178

a Write down the full structural formula of the triester made from propane-1,2,3-triol and ethanoic acid. Using a model-making kit make a model of this simple triester.

b Write equations to show what happens when this ester is hydrolysed, i.e. heated with hot concentrated sodium hydroxide solution and then neutralised by the addition of hydrochloric acid.

A selection of fats and oils.

Unlike the simple triester you constructed in the last exercise, the triesters found in natural oils and fats are often mixed triesters in which the three acid groups are different. In addition the alkyl groups frequently contain between 12 and 18 carbon atoms and may be saturated or unsaturated. Figure 10 shows the structure of a naturally occurring oil. Like all triesters it can be broken down in the body by enzymes to produce the parent propane-1,2,3-triol and carboxylic acids (fatty acids).

Figure 10
A natural fat molecule.

$$
\begin{array}{c}
H \\
| \quad\quad\quad O \\
| \quad\quad\quad \| \\
H\!-\!C\!-\!O\!-\!C\!-\!(CH_2)_7CH\!=\!CH(CH_2)_7CH_3 \\
| \quad\quad\quad O \\
| \quad\quad\quad \| \\
H\!-\!C\!-\!O\!-\!C\!-\!(CH_2)_7CH\!=\!CHCH_2CH\!=\!CH(CH_2)_4CH_3 \\
| \quad\quad\quad O \\
| \quad\quad\quad \| \\
H\!-\!C\!-\!O\!-\!C\!-\!(CH_2)_7CH\!=\!CH(CH_2)_7CH_3 \\
| \\
H
\end{array}
$$

Fatty acid side-chains

Fish oil, a rich source of arachidonic acid.

Table 8 lists data on some naturally occurring fatty acids. As well as being a good source of energy some fatty acids are vital for body processes. For example, deficiencies in linoleic acid and arachidonic acids results in skin complaints. The role of evening primrose oil in the treatment of eczema may be due to its content of linolenic acid. We look at lipids and health later, but first we see how the structures of different fats affect their physical properties. You use some of the information in Table 8 to answer the next exercise.

Table 8 Structure of some common fatty acids

No. of carbon atoms	Trivial name	Systematic name	Source	Structural formula	M.p./°C	Skeletal formula
12	Lauric acid	Dodecanoic acid	Coconut oil	$CH_3(CH_2)_{10}COOH$	44.2	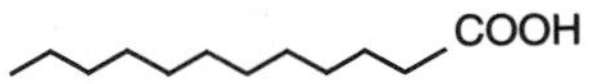
14	Myristic acid	Tetradecanoic acid	Nutmeg seed fat	$CH_3(CH_2)_{12}COOH$	53.9	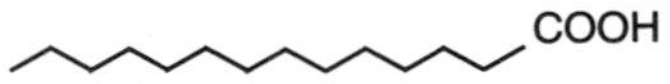
16	Palmitic acid	Hexadecanoic acid	Palm oil	$CH_3(CH_2)_{14}COOH$	63.1	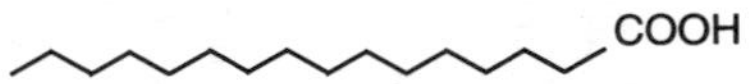
18	Stearic acid	Octadecanoic acid	Animal fats	$CH_3(CH_2)_{16}COOH$	69.6	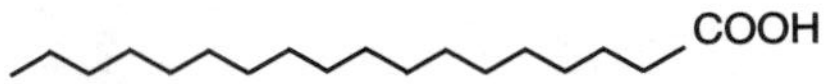
16	Palmitoleic acid	Hexadec-*cis*-9-enoic acid	Fish oil	$CH_3(CH_2)_5CH = CH(CH_2)_7COOH$	0.5	
18	Elaidic acid	Octadec-*trans*-9-enoic acid	Major constituent of partially hydrogenated fats/oils in margarines	$CH_3(CH_2)_7CH = CH(CH_2)_7COOH$	45.5	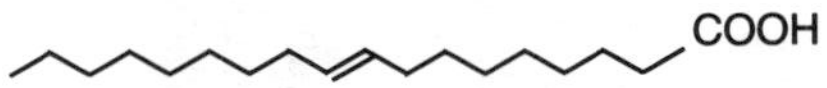
18	Oleic acid	Octadec-*cis*-9-enoic acid	Olive oil	$CH_3(CH_2)_7CH = CH(CH_2)_7COOH$	13.4	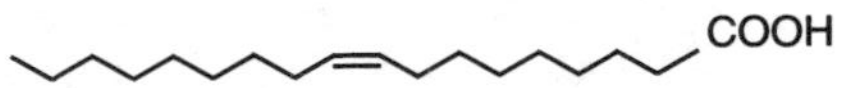
18	Linoleic acid	Octadec-*cis*-9,*cis*-12-dienoic acid	Soya bean oil	$CH_3(CH_2)_4CH = CHCH_2CH = CH(CH_2)_7COOH$	−5	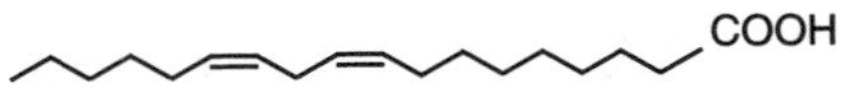
18	Linolenic acid	Octadec-*cis*-9,*cis*-12,*cis*-15-trienoic acid	Linseed oil, evening primrose oil	$CH_3CH_2CH = CHCH_2CH = CHCH_2CH = CH(CH_2)_7COOH$	−11	
20	Arachidonic acid	Eicosa-*cis*-5,*cis*-8,*cis*-11,*cis*-14-tetra-enoic acid	Fish oil	$CH_3(CH_2)_4CH = CHCH_2CH = CHCH_2CH = CHCH_2CH = CH(CH_2)_3COOH$	−49.5	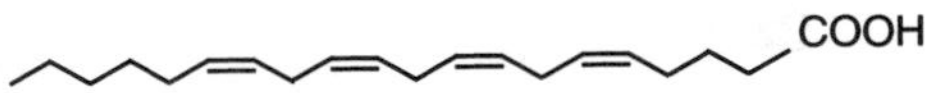

EXERCISE 36
Answers on page 178

a Name the fatty acids that will be produced in the body by the action of enzymes on the oil molecule shown in Fig. 10.
b Structural differences between the fatty acids shown in Table 8 allows us to classify them as saturated, mono-unsaturated or polyunsaturated acids. Using examples from Table 8 explain what these terms mean.

In general, animal fats (apart from fish oils) are more saturated than vegetable oils. Most unsaturated fats and oils contain fatty acid groups in the *cis* conformation. In Table 8 the molecules of fatty acids appear as linear (zigzag) chains. What is not shown is the kink that arises in unsaturated fatty acids in the *cis* conformation. This is, however, shown clearly in Fig. 11 where we compare the shape of a saturated fatty acid with a *cis* and *trans* unsaturated fatty acid.

Figure 11

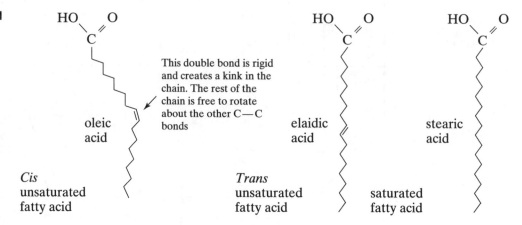

oleic acid

This double bond is rigid and creates a kink in the chain. The rest of the chain is free to rotate about the other C—C bonds

elaidic acid

stearic acid

Cis unsaturated fatty acid

Trans unsaturated fatty acid

saturated fatty acid

 In order to appreciate the difference in shape between an unsaturated fatty acid in the *cis* and *trans* conformation and a saturated fatty acid you should ask for a model-making kit and make models of the three molecules shown in Fig. 11. In view of the large number of atoms involved it is probably best performed as a group activity. You should then be able to do the next exercise.

EXERCISE 37
Answers on page 178

Using Table 8 and Fig. 11 as a guide, explain:
a What part the length of the side-chain fatty acid group plays in the melting points of fats.
b The difference in melting point between saturated and unsaturated fats with fatty acid groups in the *cis* conformation.
c The difference in melting point between unsaturated fats with fatty acid groups in the *cis* and *trans* conformation.

Vegetable oils are of little use for spreading on bread. In the next section you see how they can be converted to a solid fat, i.e. margarine.

■ 3.2 Hydrogenation of fats (margarine)

Vegetable oils can be converted to margarine by saturating some of the fatty acid side groups as you see in the next exercise.

EXERCISE 38

Answers on page 179

A food company wants to convert a vegetable oil into margarine by treating it with hydrogen in the presence of a metal catalyst.

a Give the conditions for hydrogenation and the name of a suitable catalyst.

b Complete the following equation for the addition of hydrogen across the carbon–carbon double bond in one of the side-chain fatty acid groups of a vegetable oil.

$$-CH_2 \diagdown \diagup CH_2- $$
$$C=C + H_2 \longrightarrow$$
$$H \diagup \diagdown H$$

c Explain why this process makes the oil more solid.

Not all the bonds will be hydrogenated by the hardening process; only some degree of saturation is required. By controlling the conditions, the manufacturer can saturate the required number of carbon–carbon double bonds and so produce a spread of the desired texture.

Most natural fats and oils contain only *cis* alkenes in the fatty acid groups. *Trans* fatty acid groups are found in small amounts in dairy products but much larger amounts are found in some of the harder margarines. The undesirable effects of *trans* fatty acids are considered in Section 3.4.

In the next exercise you consider how, in the process of hydrogenation, the switch from the desirable *cis* to the undesirable *trans* fatty acids might come about.

EXERCISE 39

Answers on page 179

Part of an unsaturated fat molecule in the *cis* form is shown below. When this oil is converted into margarine in the process of hydrogenation some of the *cis* forms like this are converted into *trans* forms **without** addition of hydrogen.

$$-CH_2 \diagdown_{} \diagup CH_2-$$
$$C = C \longrightarrow$$
$$H \diagup \diagdown H$$
$$cis$$

a Draw the *trans* form for this portion of the fat molecule.

b During hydrogenation what must the catalyst do to the π bond in the *cis* form before isomerisation into the *trans* form can take place?

To check that the required number of carbon–carbon double bonds have been saturated it is important to know the degree of unsaturation in the oil before and after dehydrogenation. This is now done by chromatography but was originally done by reacting the fat with iodine to obtain a value called iodine number (value).

■ 3.3 Iodine number (or iodine value)

This old method for determining the degree of unsaturation in a fat is still used by most students following a course of food science and will be understood after you have attempted the next exercise which gives you the results of an experiment to process.

EXERCISE 40

Answers on page 179

Iodine monochloride, ICl, is used to determine the degree of unsaturation in oils. The ICl adds rapidly to the carbon–carbon double bonds present. In an experiment, 0.127 g of an unsaturated oil was treated with 25.0 cm^3 of 0.100 M iodine monochloride solution. The mixture was kept in the dark until the reaction was complete. The unreacted ICl was then treated with an excess of aqueous potassium iodide, forming I_2. The liberated iodine was found to react with 40.0 cm^3 of 0.100 M sodium thiosulphate.

a Suggest why it is necessary to keep the mixture of oil and iodine monochloride in the dark.

b Write an equation for the reaction between iodine monochloride and potassium iodide.

c Calculate the number of moles of sodium thiosulphate which were used in the titration.

d Calculate the number of moles of iodine liberated, given that iodine reacts with sodium thiosulphate according to the equation:

$$I_2 + 2Na_2S_2O_3 \rightarrow 2NaI + Na_2S_4O_6$$

Hence, calculate the number of moles of unreacted iodine monochloride.

e Calculate the number of moles of iodine monochloride that reacted with the 0.127 g of unsaturated oil.

f Direct addition of iodine to an unsaturated oil is slower than the addition of ICl. However, unsaturation is quoted as the **iodine number**. The iodine number is the number of grams of iodine that in theory can be added to 100 g of oil. Calculate the iodine number of this oil, given that one mole of ICl is equivalent to one mole of I_2.

A measure of the overall unsaturation of a fat is therefore given by its iodine number. Clearly polyunsaturated fats will have high iodine numbers and saturated fats will have an iodine value of zero. Table 9 gives some approximate iodine numbers for a selection of fats and oils.

Table 9

Oil	Iodine number
Soya	130
Sunflower	120
Groundnut	90
Olive	80
Lard	60
Butter	30

Note the lower iodine number for olive oil compared with sunflower oil is due to a higher proportion of monounsaturates in the former and a higher proportion of polyunsaturates in the latter.

Current scientific evidence suggests that *trans* fats behave in a similar way to saturated fats in that they tend to raise the level of cholesterol in the blood. We consider the possible causes and consequences of this in the next section.

■ 3.4 Fats, cholesterol and health

OBJECTIVE When you have finished this section you should be able to:
■ consider present day thinking about the connection between **fatty acids and cholesterol in diet and health**.

You may find it difficult to find up-to-date information about the role of fatty acids and cholesterol in our diet and how it affects our health. Conflicting advice has been given in the past and it is hardly surprising that the consumer is left in some doubt about which type of fat to eat. Your teacher may be able to provide you with articles from scientific journals or magazines. Alternatively, we have set out an account which forms part of a comprehension exercise.

Lipids – a consumer's guide
A major contributor to coronary heart disease is a high blood cholesterol level, which in turn is determined to a large extent by the type of fat we eat. Cholesterol, like fat, is a lipid but it belongs to a separate group of compounds called steroids. It is essential for life, playing a crucial role in making certain hormones. Most cholesterol in the blood is made by the liver from a variety of foods but especially saturated fats. It is transported round the body attached to two types of lipoproteins called high-density lipoprotein (HDL) which acts like a sponge and rids the body of cholesterol, and low-density lipoprotein (LDL) which carries most of the cholesterol round the body.

If the body is unable to get rid of high cholesterol levels, this soft, waxy substance builds up in the artery walls making them narrower, slowing down blood flow or even cutting off supply to the heart. A moderate cholesterol level is 5.2–6.5 mmol/l. A cholesterol level above 6.5 mmol/l is considered too high and action should be taken to reduce it. These levels refer to the total cholesterol, that is: a mixture of LDL-cholesterol (bad cholesterol) and HDL-cholesterol (good cholesterol). So, not all cholesterol is bad for us – to quote Professor Brian Pentecost, medical director of the British Heart Foundation, 'It is important that we don't see cholesterol as a poison because we do need a certain amount of it. What matters is the type and level of the cholesterol'.

Figure 12
Cholesterol – found in many
membranes.

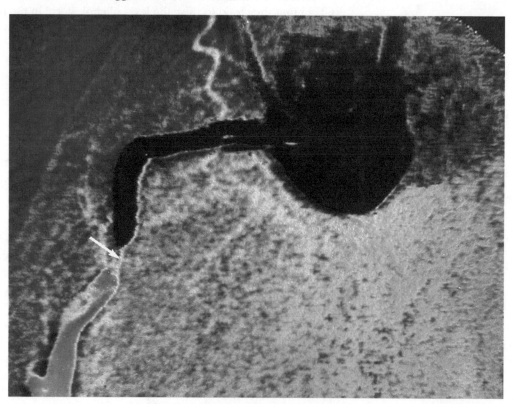

EXERCISE 41
Answers on page 180

a Why are high cholesterol levels associated with an increased risk of dying from heart disease?
b Why are total cholesterol levels misleading?
c What can be done to lower cholesterol levels?
d There is increasing evidence that HDL levels show an inverse correlation to coronary heart disease. Suggest a reason for this.

A coronary angiogram
showing a severe blockage
(see arrow).

At the time of writing, the advice given by various groups of scientists is to lower our total fat intake, particularly saturated fats, and avoid a high intake of *trans* fatty acids (mainly from hydrogenated fat in some margarines and shortening and products from them e.g. biscuits and pastries). Furthermore the British Heart Foundation reinforces this by stating that people should eat less fat overall rather than simply switching from one sort to another. With this in mind you should attempt the next group discussion. Extracts from a Department of Health report (1994) 46 *Nutritional Aspects of Cardiovascular Disease* giving recommendations on fats in our diet is included in the ILPAC Teacher's and Technician's Notes.

EXERCISE

Using the information in this section you should hold a group discussion and decide which fats and oils you might try to cut down on or avoid for a healthier diet.
You should now attempt the following exercise.

EXERCISE 42

Answers on page 180

Food analysts often represent fatty acids in shorthand form according to the number of carbon atoms and the number of carbon–carbon double bonds in the molecule, for example, stearic acid is C 18:0 and linoleic acid is C 18:2.

a On the basis of this information, draw a possible displayed formula (showing all atoms and bonds) for linoleic acid.

b State whether stearic acid and linoleic acid are saturated or unsaturated fatty acids. Justify your answers.

c Draw the displayed formula of the triglyceride formed between glycerol (propane-1,2,3-triol) and stearic acid. (The molecular formula for the carbon chain can be shown as $C_{17}H_{35}$.)

d Table 10 compares the fatty acid composition of palm, sunflower seed and rape seed oils for five major fatty acids.

Table 10

| | Fatty acid composition/% by mass of total fat | | | | |
	C16:0	C18:0	C18:1	C18:2	C18:3
Palm oil	44.1	4.4	40.1	10.1	0.3
Sunflower seed oil	6.3	4.8	24.2	64.4	0.1
Rape seed oil	4.7	1.8	59.9	20.8	10.3

By reference to Table 10, answer the following questions.

i)   Which oil has the lowest proportion of saturated fatty acids?

ii) Which oil has the highest proportion of monounsaturated fatty acids?

iii) Which oil has the highest level of polyunsaturated fatty acids?

iv) Which oil is best avoided by a person who has suffered a mild heart attack? Justify your answer.

The sodium and potassium salts of long-chain carboxylic acids (fatty acids) have been used for centuries as soaps to make washing more effective. More recently, with the growth of the petrochemicals industry, synthetic detergents have been developed which are superior in some respects to soap. We look at both types in the next section. If this topic is not a requirement of your particular syllabus skip this section and proceed to page 55.

■ 3.5 Soaps and detergents

In this section we consider the structures of soap and detergent molecules and the way in which they act as wetting and grease-removing agents.

OBJECTIVES

When you have finished this section you should be able to:

■ write down the general structure of a **soap** (or soapy detergent) and a **soapless detergent**;

■ explain how soaps and detergents act as **wetting** and **grease-removing agents**.

Read about soapless detergents in your textbooks. Look for their general formulae, noting the difference between anionic and cationic detergents. Also look for their advantages over soap, and the way they work as grease-removing agents and wetting agents. Other important points are the difference between biodegradable and non-biodegradable, and biological and non-biological detergents. You should then be able to do the following exercises.

EXERCISE 43

Answer on page 180

Ordinary soaps are sodium salts of long-chain carboxylic acids (fatty acids). Their general formula is $R—COO^-Na^+$, where R is a long hydrocarbon chain. Write an equation to show how you could produce the most common soap, sodium stearate, from animal fat.

A selection of solid soaps.

Because of some disadvantages with the use of soap, scientists developed a new type of cleaning agent – soapless detergents which, instead of using animal fats, use products from crude oil. The early synthetic detergents were non-biodegradable and polluted rivers with foam. This problem has largely been overcome by altering the hydrocarbon 'tails' of the detergent molecules.

EXERCISE 44

Answers on page 181

a Products such as the molecule shown are extremely useful as detergents.

$$C_9H_{19}—\bigcirc—SO_2O^-Na^+$$

What advantages do these substances have over soap?

b How do the hydrocarbon 'tails' of biodegradable and non-biodegradable detergents differ?

A selection of soapless detergents.

EXERCISE 45

Answers on page 181

Figure 13 shows a simplified diagram of a soap or detergent molecule. Figure 14 illustrates how these molecules reduce the surface tension of water and finally, Fig. 15 shows how soaps and detergents separate grease from cloth. Study each in turn and then attempt the questions that follow.

Figure 13
Simplified way of showing a detergent molecule.

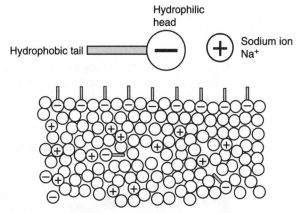

Figure 14
How detergents reduce the surface tension of water.

Figure 15
How detergents remove grease from cloth. (Water molecules not shown)

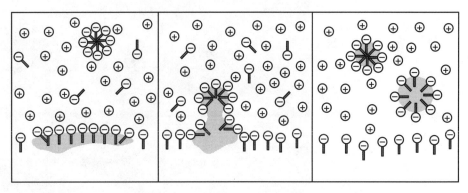

Make sketches of Figs. 14 and 15 and provide your own labels and captions to describe the action of detergents in:
a lowering the surface tension of water,
b the removal of fats and oils.

Take a look at Fig. 16 which shows the contents listed on a box of washing powder. You consider the function of some of these products in the next exercises.

Figure 16
You can see that the one on the right is a biological detergent and lists enzymes in its ingredients.

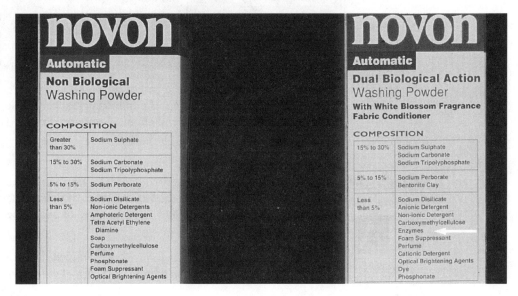

EXERCISE 46
Answers on page 182

a Describe each of the following as either anionic or cationic detergents:

i) $RSO_2O^-Na^+$
ii) $RN^+(CH_3)_3Cl^-$
iii) $ROSO_3^-Na^+$

iv) R—⬡—$SO_2O^-Na^+$

v) R_3N^+—⬡ Cl^-

b Although detergents are better than soap in hard water, they are still affected by Mg^{2+} and Ca^{2+} ions to a certain extent. Explain how the addition of phosphates could improve the function of detergents in hard water. (See ILPAC 12, page 31.)

c The addition of phosphates to detergents has been blamed for making some waterways stagnant and lifeless. Explain how this has come about.

d Why are lower temperatures generally recommended when using biological washing powders?

Algal bloom at a water treatment works, Ormesby, Norfolk.

Now that you have reached the end of the section on detergents you should look back through your notes and attempt the following teacher-marked exercise. Make a plan of your answer first and then spend about 35 minutes writing. Hand both your plan and the answer in to your teacher for marking.

EXERCISE
Teacher-marked

A

What is understood by the word 'detergent'? Explain in detail the difference in molecular structure between the various types of detergent, both soapy and soapless, and mention the advantages and disadvantages of each type.

Another use for fats and oils is to provide carboxylic acids as starting materials for the synthesis of long-chain carbon compounds with eight or more carbon atoms.

■ 3.6 Fats and oils in synthesis

OBJECTIVE When you have finished this section you should be able to:
■ suggest some synthetic pathways starting with a named fat or oil.

The synthesis of compounds with an even number of carbon atoms is straightforward because, as you can see from Table 8, in most fats and oils the carboxylic acids from which they are derived have an even number of carbon atoms. The synthesis of compounds with an odd number of carbon atoms is more costly because it requires the introduction or removal of one carbon atom. You suggest some synthetic pathways starting from fats and oils in the next teacher-marked exercise.

EXERCISE
Teacher-marked

Fats and oils are suitable sources of long-chain carbon compounds. Those that are esters are hydrolysed by mineral acids to produce propane-1,2,3-triol and a carboxylic acid with an even number of carbon atoms.

Outline, in a reaction sequence with names and formulae, how you could make each of the following substances starting from a named fat or oil. You should include reagents and conditions but balanced equations are not required.
a Tetradecanal, $C_{13}H_{27}CHO$.
b Octan-2-ol, $C_6H_{13}CH(OH)CH_3$.
c Octadecane-1,9,10-triol, $C_8H_{17}CH(OH)CH(OH)C_7H_{14}CH_2OH$.

[Hint: Table 8, page 45, gives a selection of oils and fats and the carboxylic acids which they would produce on hydrolysis.]

When crude oil runs out we could be looking to vegetable oils not only as a feed-stock for the chemicals industry but the fuels for industry as well. Already, a bus company in Reading, UK, has run some of its vehicles on modified rapeseed oil. Vegetable oil itself is not volatile enough for use as a petrol substitute. Fuel technologists can turn the oil into small gasoline-like molecules with the aid of catalytic cracking. You consider the advantages and disadvantages of vegetable oils as fuels in the next teacher-marked exercise. You may wish to refer back to Exercise 40 for the second part of the question.

EXERCISE
Teacher-marked

Most modern vehicles run on petrol or diesel fuel, both of which are mainly a mixture of compounds from the homologous series of alkanes. One surprising alternative to diesel fuel is rapeseed oil mixed with ethanol. This is used on a small scale by farmers in their tractors. Rapeseed oil is advertised as being high in unsaturated compounds. It is easily obtained by crushing the seed followed by filtration.

a Discuss, in economic and environmental terms, the advantages and disadvantages of using rapeseed oil instead of diesel as a tractor fuel.
b i) Devise a simple laboratory experiment to compare the extent of unsaturation of two oils.
   ii) Indicate how such oils could be saturated.

Rapeseed oil can be used as
a fuel for tractors.

Carboxyl groups are to be found along with another functional group – amino, NH_2 (see
ILPAC 8, Functional Groups) in amino acids and proteins. These form the subject of the
next chapter. Since syllabus requirements vary considerably on this particular topic you
must check with your teacher or syllabus to find out which parts apply to you. Some of
the extension material involving the determination of protein structure, for instance, is
more relevant for those studying special options or modules in biochemistry or food
science.

CHAPTER 4

AMINO ACIDS

Proteins are often called natural polymers but, unlike synthetic ones, they are made up from a selection of about twenty different sub-units, all of which are compounds known as amino acids. Clearly, before you can study proteins, you need to know what amino acids are and how they react.

■ 4.1 Formulae of amino acids

In this section you examine the formulae of a number of naturally occurring amino acids to establish some common features. You also revise some of what you have learned about systematic names.

OBJECTIVES When you have finished this section you should be able to:
- write the **general formula** for an **α-amino acid**;
- state the systematic (IUPAC) name for some amino acids;
- recognise an **optically active** amino acid from its formula.

 Read the introductory section on amino acids in your textbook(s). Look for a generalised formula and find out what are the distinguishing features of naturally occurring amino acids.

Table 11 shows the names and formulae of some of the common amino acids. Don't worry – you need not learn **all** of these! It will probably be sufficient for you to be able to quote glycine and alanine as examples unless you are studying biochemistry as a special option, in which case your teacher will suggest what else you need to know.

Notice that the trivial names of amino acids, often in an abbreviated three-letter form, are still widely used because some of the systematic names are very clumsy. In the next exercise, you try naming some amino acids systematically.

EXERCISE 47
Answers on page 182

a What structural features are common to all the molecules shown in Table 11?
b What is the significance of the letter α in the title α-amino acid?
c Write systematic names for:
  i) glycine,   iii) serine,   v) lysine,
  ii) alanine,   iv) phenylalanine,   vi) valine.
d Why are systematic names not often used for amino acids?
e Which of the amino acids in Table 11 would you **not** expect to be optically active?

We now consider the behaviour of amino acids, beginning with physical properties and acid–base character.

■ 4.2 Physical properties and acid–base behaviour of amino acids

OBJECTIVES When you have finished this section you should be able to:
- explain how the **physical properties of amino acids** provide evidence for the existence of **zwitterions**;
- write equations for the **reactions of an amino acid** with acids and alkalis;
- distinguish between **acidic**, **basic** and **neutral amino acids**;
- explain the term **isoelectric point** and show how amino acid solutions act as **buffers**;
- outline the process of **DNA fingerprinting**.
- describe simply the process of **electrophoresis** in the separation of a mixture of amino acids.

Table 11
Naturally occurring amino acids

[1] There is also an amide form called asparagine, asn ($-CONH_2$ for $-CO_2H$).

[2] Pronounced 'sistane'. Not to be confused with cystine ('sisteen'), cys–cys, which consists of two cys units linked through the S atoms.

[3] There is also an amide form called glutamine, gln ($-CONH_2$ for $-CO_2H$).

[4] Hydroxylysine has an $-OH$ group substituted in the C_5 position shown.

[5] Hydroxyproline has an $-OH$ group substituted in the C_4 position shown.

Name and abbreviation		Formula
Alanine	ala	H_2NCHCO_2H $\|$ CH_3
Arginine	arg	H_2NCHCO_2H $\|$ $(CH_2)_3$ $\|$ NH $\|$ $HN=C-NH_2$
Aspartic acid[1]	asp	H_2NCHCO_2H $\|$ CH_2 $\|$ CO_2H
Cysteine[2]	cys	H_2NCHCO_2H $\|$ CH_2 $\|$ SH
Glutamic acid[3]	glu	H_2NCHCO_2H $\|$ CH_2 $\|$ CH_2 $\|$ CO_2H
Glycine	gly	H_2NCHCO_2H $\|$ H
Histidine	his	H_2NCHCO_2H $\|$ CH_2 (imidazole ring with NH and N)
Isoleucine	ile	H_2NCHCO_2H $\|$ $CH-CH_3$ $\|$ CH_2 $\|$ CH_3
Leucine	leu	H_2NCHCO_2H $\|$ CH_2 $\|$ $CH_3-CH-CH_3$

Name and abbreviation		Formula
Lysine[4]	lys	H_2NCHCO_2H $\|$ $(CH_2)_4$* $\|$ NH_2
Methionine	met	H_2NCHCO_2H $\|$ $(CH_2)_2$ $\|$ S $\|$ CH_3
Phenylalanine	phe	H_2NCHCO_2H $\|$ CH_2 (benzene ring)
Proline[5]	pro	$HN-CHCO_2H$ (ring) *
Serine	ser	H_2NCHCO_2H $\|$ CH_2OH
Threonine	thr	H_2NCHCO_2H $\|$ $CHOH$ $\|$ CH_3
Tryptophan	trp	H_2NCHCO_2H $\|$ CH_2 (indole ring with NH)
Tyrosine	tyr	H_2NCHCO_2H $\|$ CH_2 (benzene ring) $\|$ OH
Valine	val	H_2NCHCO_2H $\|$ $CH_3-CH-CH_3$

Before you start work on this section, you may need to refresh your memory on the reactions of the carboxyl group you studied earlier in this book and read through your notes from ILPAC 7, Equilibrium II: Acids and Bases, on the properties of buffer solutions and ILPAC 8, Functional Groups, to revise the reactions of the amino group. Read about the physical properties of amino acids, looking for an explanation of the term 'zwitterion' (or **'dipolar ion'**). This will enable you to do the next exercise. Also read about the technique called electrophoresis used to separate mixtures of amino acids.

EXERCISE 48
Answer on page 182

Why does glycine have such a high melting point compared with the compounds of similar molar mass listed in Table 12?

Table 12

Compound	Molar mass/g mol^{-1}	Melting point/°C
Glycine	75	235
Propanoic acid	74	−21
Methyl ethanoate	74.1	−98
1-aminobutane	73.1	−49

Amino acids are bifunctional compounds containing both amino and carboxyl groups. We would therefore expect them to react with both acids and alkalis, an idea that you explore in the next exercise.

EXERCISE 49
Answer on page 182

Write equations to show how you would expect a molecule of alanine, in the form of $CH_3CH(NH_2)CO_2H$, to react with:
a dilute aqueous sodium hydroxide,
b dilute hydrochloric acid.

When an amino acid is dissolved in water, a series of equilibria is set up, as you find out in the next exercise.

EXERCISE 50
Answer on page 183

The equations below show the equilibria that are set up when the amino acid glycine is in solution:

$$H_2N-CH_2-CO_2^- \underset{\pm H^+}{\rightleftharpoons} H_3N^+-CH_2-CO_2^- \underset{\pm H^+}{\rightleftharpoons} H_3N^+-CH_2-CO_2H$$
$$\mathbf{A} \qquad\qquad \mathbf{B} \qquad\qquad \mathbf{C}$$

a Which of the forms **A**, **B** and **C** would you expect to be in greatest concentration at low pH and at high pH? Explain.
b Describe the forms **A**, **B** and **C** using the terms conjugate acid, conjugate base and zwitterion.
c Which of the three forms do you think would show no net movement in an electric field?

The pH at which the zwitterion form is at maximum concentration is the **isoelectric point** or **isoelectric pH**. At this pH, an amino acid possesses no net charge and does not migrate in an electric field. This is useful in the separation and identification of amino acids by the technique of electrophoresis. With this in mind, attempt the next two exercises.

EXERCISE 51
Answer on page 183

Table 13

Amino acid	Formula	Isoelectric point
Arginine	$HN = C(NH_2)NH(CH_2)_3CH(NH_2)CO_2H$	10.8
Aspartic acid	$CO_2HCH_2CH(NH_2)CO_2H$	2.8
Glutamic acid	$CO_2H(CH_2)_2CH(NH_2)CO_2H$	3.2
Glycine	$CH_2(NH_2)CO_2H$	6.0
Leucine	$(CH_3)_2CHCH_2CH(NH_2)CO_2H$	6.0
Lysine	$H_2N(CH_2)_4CH(NH_2)CO_2H$	10.0

a By examining the formulae of the amino acids shown in Table 13, classify them as acidic, basic or neutral.

b Is there a relation between isoelectric point and the type of amino acid?

The fact that amino acids and proteins have different isoelectric points (isoelectric pH) is made use of in a technique designed to separate them from mixtures. This technique is called electrophoresis. Check whether this is a requirement of your syllabus.

■ 4.3 Electrophoresis

Figure 17 shows a schematic diagram of the apparatus used to separate amino acids or proteins by electrophoresis. Study it together with your answer to Exercise 51 and then attempt the next exercise.

Figure 17

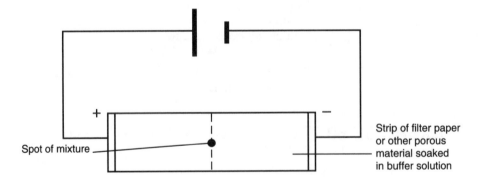

Spot of mixture

Strip of filter paper or other porous material soaked in buffer solution

EXERCISE 52
Answers on page 183

a If the isoelectric point (pH) of an amino acid in a mixture is:
  i) lower than the pH of the buffer solution,
  ii) higher than the pH of the buffer solution,
  will it move towards the anode or cathode? Explain.

b What will happen to the amino acids whose isoelectric point (pH) is the same as the buffer solution?

c What factors will affect the rate of migration of a particular amino acid?

So the migration at a particular pH will enable a particular amino acid or protein to be identified. This method is frequently used in hospital pathology laboratories where analysis of blood proteins can help diagnose a number of diseases.

One recent application of electrophoresis is in DNA fingerprinting which is the topic of the next section. A section on nucleic acids (DNA) is included in Appendix 2. Check whether this is a requirement of your particular syllabus; if not, proceed to Section 4.5.

■ 4.4 DNA fingerprinting

You have probably heard of cases where forensic scientists have produced evidence using the technique of DNA fingerprinting to convict or to clear a suspect of a particular crime. The technique was discovered by accident in 1983 during research by Dr Alec Jeffreys, a geneticist at Leicester University.

If possible, read about DNA fingerprinting (your teacher will have the reference). SATIS 16–19 Unit 6 is also very useful. 'DNA' *Education Guardian*, 10 September 1991 has a very good article on this. Your school librarian or teacher may have back copies of this very useful resource. Look out for a simple description of the stages involved in the process of analysis of genes and genetic fingerprinting and how the technique can be used to establish family relationships.

Examining DNA fingerprints.

EXERCISE 53
Answers on page 183

The following stages are involved in the process of DNA fingerprinting. Arrange them in the correct sequence.

A. DNA molecules are cut into fragments by enzymes.
B. The exposed areas on the developed film form bands which make up the DNA fingerprint.
C. Blood sample is taken.
D. Radiation from the nylon membrane produces a pattern on a sheet of X-ray film.
E. DNA molecules are extracted from the white blood cells.
F. Radioactive DNA probes are allowed to bind to specific DNA sequences on the nylon membrane with its DNA pattern.
G. DNA fragments are separated by electrophoresis along a sheet of gel.
H. DNA fragments are transferred from the gel to a nylon membrane.

In the process of DNA fingerprinting a photographic film is produced showing DNA fragments which have been separated by electrophoresis. It looks rather like a bar code you see on shop goods. Samples of blood, semen and hair root produce DNA fingerprints which are specific to individuals. Since half the DNA comes from the mother and half from the father there are similarities between parents and children. With this in mind attempt the next exercise.

EXERCISE 54
Answer on page 183

Figure 18 shows a diagrammatic representation of the DNA fingerprint of a mother and son. The other two DNA fingerprints are produced from two men, A and B. Study and compare each and then decide which of the men, A or B, is the boy's biological father.

Figure 18

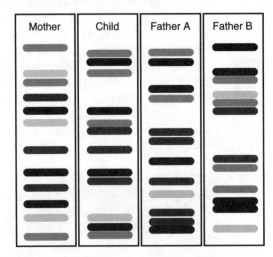

You should now be able to attempt the next teacher-marked exercise.

EXERCISE
Teacher-marked

a Polymers of biological interest are usually electrically charged and can be effectively studied by electrophoresis.
Describe this technique, including in your answer:
i) a description of the apparatus,
ii) an account of the principles involved,
iii) the effect of pH on the results.
b Outline examples of how this technique may be used for the analysis of genes* and genetic fingerprinting.

Another consequence of the presence of both acidic and basic groups is that they can be used in buffers.

■ 4.5 Buffer action of amino acids

In the next section you apply your knowledge of the ionisation of amino acids to see how they can act as buffer solutions. You may need to revise the expression you first met in ILPAC 7, Equilibrium II: Acids and Bases:

$$pH = pK_a - \log \frac{[acid]}{[base]}$$

EXERCISE 55
Answers on page 183

Figure 19 (opposite) is a titration curve for glycine. (The shape may seem unfamiliar to you because pH is shown here along the x-axis, not the y-axis as in ILPAC 7, Equilibrium II: Acids and Bases.)

a Use formulae to show which form of glycine would predominate at points **A**, **C** and **E** in Fig. 19.
b The pK_a values for glycine are $pK_1 = 2.34$, $pK_2 = 9.60$. Use these to decide the species and the ratio in which they are present at points **B** and **D** on the graph.
c Over which ranges of pH does glycine behave as:
i) a good buffer,
ii) a poor buffer? Explain.

*DNA fragments

Figure 19
Titration curve for 0.001 mol
of glycine.

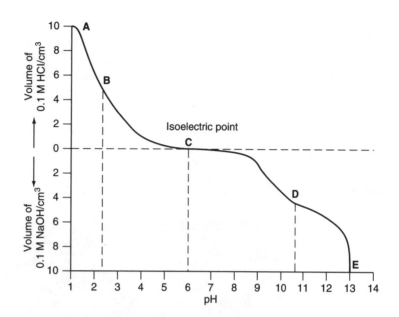

The next exercise consists of parts of two A-level questions.

EXERCISE 56

Answers on page 184

a 2-Aminopropanoic acid (alanine) has pK_a values of 2.4 and 9.7 at 298 K. Discuss the species present at different pH values when aqueous solutions of alanine at pH = 7 are titrated with:
  i) hydrochloric acid to a pH of 1.5,
  ii) aqueous sodium hydroxide to a pH of 11.5.
b State what species you think will be present in similar titrations using 2,6-diaminohexanoic acid (lysine), the pK_a values being 2.2, 9.0 and 10.5.
c The pK_a values for aspartic acid, $CO_2HCH_2CH(NH_2)CO_2H$, are 2.1, 3.9 and 9.8. From your knowledge of the acid–base behaviour of amino acids, give the structures of the species which predominate in a solution of aspartic acid at pH = 1, 7 and 13.

Now we move on to consider some further reactions of amino acids.

■ 4.6 Chemical reactions of amino acids

Most of these reactions are what you would expect from your knowledge of the two functional groups.

OBJECTIVES

When you have finished this section you should be able to:
■ state how an **amino acid reacts** with **nitrous acid, ethanol, ethanoic anhydride and soda-lime;**
■ describe the **reaction** between **aqueous glycine** and **copper(II) ions**.

Read about the chemical reactions of amino acids, comparing them with what you already know about amino compounds and carboxylic acids. You should then be able to do the following exercises.

EXERCISE 57

Answers on page 185

a Write an equation for the reaction between alanine and nitrous acid.
b i) Which reagents could you use to acylate alanine? (Acylation is the substitution of $RC = O$ for H.)
   ii) Write an equation to show acylation using one of these reagents.
c i) Write an equation for the esterification reaction between alanine and ethanol.
   ii) How is this reaction catalysed?

EXERCISE 58
Answers on page 186

Use the information below to identify, by name and formula, substances G, H and I.

G is a crystalline solid which has a high melting point (over 200 °C) and dissolves in water to give an almost neutral solution. It forms crystalline salts with both acids and bases, and yields a pungent-smelling gas, H, when heated with soda-lime. H burns in air and dissolves in water to give an alkaline solution. When G is treated with nitrous acid it yields nitrogen, water and an acid I with a relative molecular mass of 76.

Before you do the next exercise, which is about the complex formed between glycine in solution and copper(II) ions, we suggest a short experiment which should take you no more than ten minutes.

EXPERIMENT 5 **The glycine/copper(II) complex**

Aim The purpose of this simple experiment is to prepare a complex from glycine and copper(II) ions.

Requirements
- safety spectacles
- wash-bottle of distilled water
- test-tube
- spatula
- glycine, $CH_2NH_2CO_2H$
- copper(II) carbonate, $CuCO_3$ (powdered)
- stirring rod
- Büchner funnel (small) or Hirsch funnel
- filter paper
- filter tube with side-arm
- filter pump and pressure tubing
- crystallising dish

Procedure
1. To about 10 cm³ of distilled water in a test-tube, add about 0.5 g (small spatula measure) of glycine. Note how easily it dissolves.
2. Slowly add powdered copper(II) carbonate, stirring the contents of the tube between additions. Keep adding the powder until it is in excess.
3. Set up a small Büchner funnel or Hirsch funnel in a side-arm filter tube and filter the mixture.
4. Transfer the filtrate to a crystallising dish and allow it to stand. Note the colour of the solution and whether crystals are formed.

Questions
Answers on page 186

1. Copper(II) carbonate is only slightly soluble but during the experiment some CO_3^{2-} ions do dissolve. Would you expect this to make the solution slightly acidic, neutral or alkaline? Explain.
2. Bearing in mind your answer to (1), what form of glycine would you expect to predominate in the solution?
3. How many bonds is each glycine molecule capable of making with Cu^{2+}(aq) under these conditions?

EXERCISE 59
Answers on page 186

a Bearing in mind your answers to the questions following Experiment 5 and your knowledge of the types of complex formed by copper(II) ions, suggest a structural formula for the copper(II)/glycine complex.
b What shape and charge would this complex have?

Having looked at the properties of amino acids, we now turn to methods of preparation.

■ 4.7 Preparation of amino acids

There are two main methods for preparing amino acids. Each uses the reactions of functional groups that you have already studied.

OBJECTIVE
When you have finished this section you should be able to:
■ outline two general methods for the **preparation of amino acids**.

Read about the methods for making amino acids from aldehydes and via the chlorination of alkanoic acids. You may find one of these methods listed as the **Strecker synthesis**. This should enable you to do the next exercise.

EXERCISE 60
Answers on page 187

a Show how aminoethanoic acid (glycine) may be prepared, in three stages, from ethanoic acid.
b Show how 2-aminopropanoic acid (alanine) may be prepared from ethanal.
c Would you expect the products from (a) and (b) to be optically active?

To help consolidate your knowledge of the reactions of amino acids, you should now attempt the following teacher-marked exercise, which is an A-level essay-style question. Before you attempt the question, look back through your notes and make a plan of the points you think your answer should include. Spend about 40 minutes on each of your plan and your answer.

EXERCISE
Teacher-marked

Amino acids contain two different functional groups in the same molecule. Describe some typical physical and chemical properties of this class of compound and, where possible, interpret these properties in terms of the structure possessed by amino acids.

The reactions you have considered so far do not suggest that amino acids might polymerise. However, in biological systems, amino acids do link together to form compounds called peptides, polypeptides and proteins. We consider the way in which proteins are built up from amino acids in the next chapter.

PROTEINS

In biological systems, the $-NH_2$ group from one amino acid molecule is enabled to react with the $-CO_2H$ group in another molecule (or in a growing chain) by the action of a specific enzyme catalyst. Many series of such reactions, involving twenty-odd different amino acids, result in the formation of proteins, which form a large proportion of animal tissue.

In laboratory conditions it is much easier to split up proteins than it is to synthesise them. In this chapter, you see how the hydrolysis of proteins enables us to work out both the identity of the constituent amino acids and the order in which they are linked together. First, however, we look at the nature of the linkage between amino acid molecules.

■ 5.1 The peptide link

When you have finished this section you should be able to:
- ■ write a generalised equation for the formation of a **peptide link** between two amino acids;
- ■ explain why amino acids do not readily form peptide links in ordinary conditions;
- ■ explain the terms **dipeptide**, **tripeptide** and **polypeptide**.

Figure 20 shows a generalised equation for the formation of a peptide link between two amino acids. Study it and attempt the exercise that follows.

Figure 20

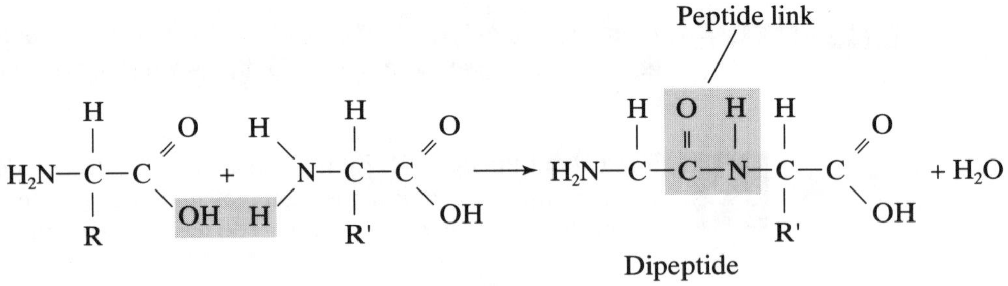

Peptide link

$$H_2N-\underset{\underset{R}{|}}{\overset{\overset{H}{|}}{C}}-C\underset{\diagdown OH}{\overset{\diagup O}{}} \quad + \quad \underset{\underset{R'}{|}}{\overset{H}{N}}-\underset{\underset{R'}{|}}{\overset{\overset{H}{|}}{C}}-C\underset{\diagdown OH}{\overset{\diagup O}{}} \longrightarrow H_2N-\underset{\underset{R}{|}}{\overset{\overset{H}{|}}{C}}-\overset{\overset{O}{\|}}{C}-\underset{\underset{R'}{|}}{\overset{\overset{H}{|}}{N}}-\underset{\underset{R'}{|}}{\overset{\overset{H}{|}}{C}}-C\underset{\diagdown OH}{\overset{\diagup O}{}} \quad + H_2O$$

Dipeptide

EXERCISE 61

Answers on page 187

a Write an equation to show a condensation reaction between the $-NH_2$ group in one molecule of alanine and the $-CO_2H$ group in another.
b From what you know of the acid–base nature of alanine, why do you think the reaction in (a) does not occur readily in aqueous solution?
c How can the alanine molecule be modified so that a similar condensation reaction does readily occur, giving the same product as in (a)?
d The product of the reaction is called alanylalanine. Why is this name used in preference to a systematic name? What **type** of compound is it?

You saw in the last exercise that it is a fairly simple matter to make a dipeptide from a single amino acid, and this dipeptide could react further in the same way to form a tripeptide and a polypeptide. This polypeptide is not, of course, a protein because it is built up from only one amino acid.

 It is not so simple to make a dipeptide from two different amino acids, A and B. In order to prevent identical molecules from combining together, the $-NH_2$ group in A and the $-CO_2H$ group in B must be protected in some way. Ask your teacher whether you should find out how this is done by reading a suitable textbook.

The mechanism of protein synthesis in biological systems is too complex for A-level study. Instead we turn directly to the structure of proteins.

■ 5.2 Structure of proteins

Early workers used the powerful technique of X-ray diffraction to unravel protein structures. We deal with the basic principles of X-ray diffraction later in this book. Check with your teacher whether you need to know this for your particular syllabus.

OBJECTIVES

When you have finished this section you should be able to:
■ distinguish between a **fibrous** and a **globular** protein;
■ explain how **restricted rotation** around the peptide bond gives it a planar shape;
■ distinguish between the **primary**, **secondary** and **tertiary** structures of a protein;
■ state and explain the roles of the **peptide bond**, **hydrogen bonding** and **disulphide bridges** in protein structure;
■ explain what happens when a protein **denatures**.

Start by reading about protein structure in a textbook. Many books classify proteins according to their biological functions, as fibrous and globular. Try to find out the structural difference between these two types. Also look for an explanation of primary, secondary and tertiary structures in proteins.

■ 5.3 Primary structure

You have already seen that a protein molecule is a long chain consisting of amino acid units joined together by peptide links. The sequence of amino acid units (sometimes called residues) is the **primary** structure of a protein.

The importance of this sequence of amino acid units is illustrated by the 'molecular disease' known as sickle cell anaemia. People suffering from this disorder have a proportion of sickle-shaped red blood cells, which absorb oxygen less efficiently than the normal disc-shaped cells (see Fig. 21).

Figure 21
Normal (disc-shaped) and sickle red blood cells.

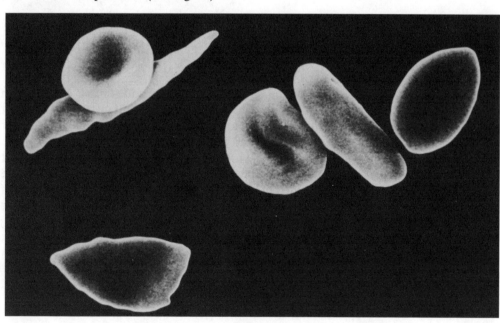

This vital difference in function has been shown to be due to the misplacement of only **one** of the 574 amino acid units in the protein haemoglobin!

Before we look at the methods by which primary structures have been determined, we consider the factors that control the shapes adopted by the long protein chains.

■ 5.4 Secondary structure

The nature of the peptide link affects the shape of a protein molecule, as you see in the next exercise.

EXERCISE 62
Answers on page 188
Figure 22

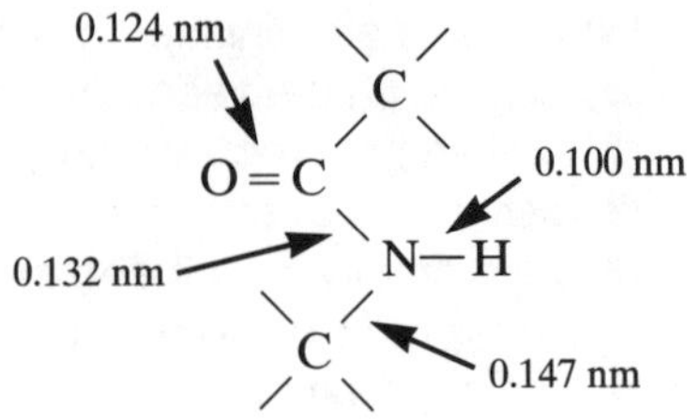

Figure 22 shows the average bond lengths for a typical peptide link.
   Average bond lengths for C—N and C=N are 0.147 nm and 0.127 nm respectively.
a Using this information, comment on the nature of the central C—N bond in a peptide link.
b Explain how the nature of this central C—N bond makes the peptide link planar.
c On a sketch outline the area that is approximately planar.

The planar peptide link has the effect of restricting the rotation at every third bond in the polypeptide chain, as shown in Fig. 23.

Figure 23

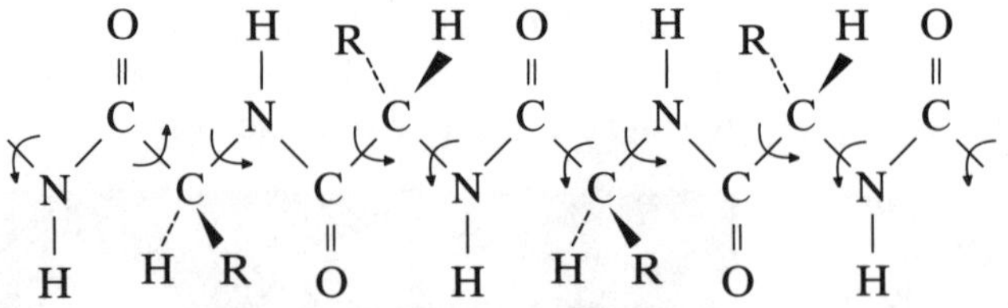

The limited rotation about the peptide links is one factor that affects the shape of the protein molecules. The most important factor, however, is the possibility of hydrogen bonding at various points within each molecule if the chain folds back on itself.
   The dimensions of the — N— C— C— N— C— C— N— C— C— N— skeleton are common to all proteins and, consequently, similar secondary structures are found in different proteins. The two most important are the β-pleated-sheet and the α-helix, illustrated in Figs. 24 and 25. In each case, large numbers of hydrogen bonds hold the different parts of the chain in fixed positions. (In Fig. 25 for the sake of clarity, the atoms constituting the **rear** part of the spiral chain are not shown.)

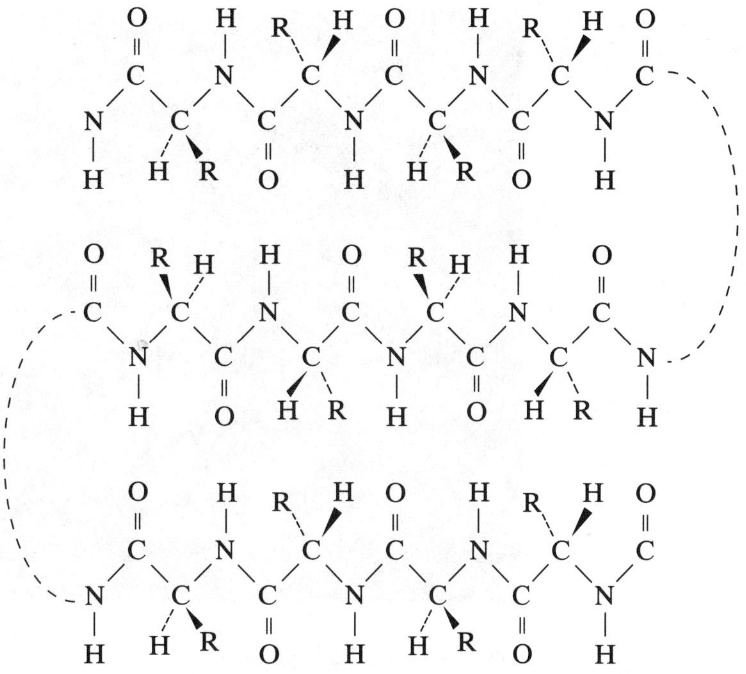

Figure 24
β-pleated-sheet structure of a protein.

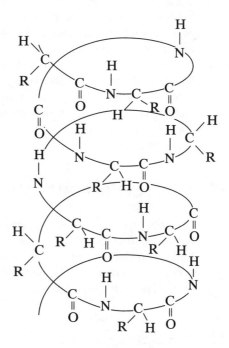

Figure 25
α-helix structure of a protein.

The next exercise concerns Figs. 24 and 25.

EXERCISE 63

Answers on page 188

a On a photocopy of Figs. 24 and 25, sketch in the hydrogen bonds.

b How would you describe the position of the R groups, relative to the peptide links in each structure?

There is also a third type of secondary structure in which two or more protein chains are twisted together like a rope. Again, hydrogen bonds hold the chains together. Proteins such as collagen, which are the tissue of muscles, tendons and ligaments, have this type of secondary structure.

Hair gets shorter and springier after blowdrying.

The nature of the R groups attached to the $-N-C-C-N-C-C-N-$ skeleton decides which type of secondary structure is adopted by a particular protein. However, some proteins change from one structure to another in different conditions. For instance, keratin, the protein in hair, usually exists as α-helices but, when it is stretched or in contact with hot water, it forms β-pleated-sheets. This is why your hair gets shorter and springier when you dry it.

Fibrous proteins, such as those in hair, silk, wool, etc., contain relatively few different amino acids, and these are arranged in regularly repeating units. Consequently, the same secondary structure is repeated throughout. In more complicated proteins, parts of the chain may be helical, but the secondary structure is folded back on itself to give a convoluted globular shape – hence the name globular proteins. We discuss an example in the next section.

■ 5.5 Tertiary structure

The folding of the polymer chain in globular proteins was once thought to be random, but X-ray diffraction has shown that each one has a definite structure. The complex folding produces reactive sites of precise dimensions to enable specific reactions to occur. For example, certain enzymes are able to promote the formation of peptide links in protein synthesis, as shown diagrammatically in Fig. 26.

Figure 26
Formation of a peptide link catalysed by an enzyme.

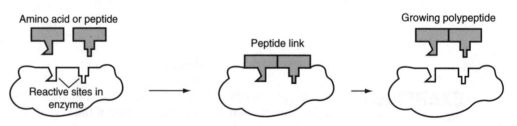

Once again, hydrogen bonding contributes to tertiary structure. Other factors are ionic bonding between acidic and basic amino acids, and sulphur bridging between cysteine units. This bridging is sometimes regarded as part of the primary structure, with cys–cys (cysteine) being regarded as a single amino acid. (See Table 11 on page 58.)

In the next exercise you explore the tertiary structure of lysozyme with the help of Figs. 27 and 28. Figure 27 shows the primary structure giving the amino acid sequence, and Fig. 28 gives an indication of the secondary and tertiary structures. You may also need your textbook.

Figure 27
Primary structure of lysozyme.

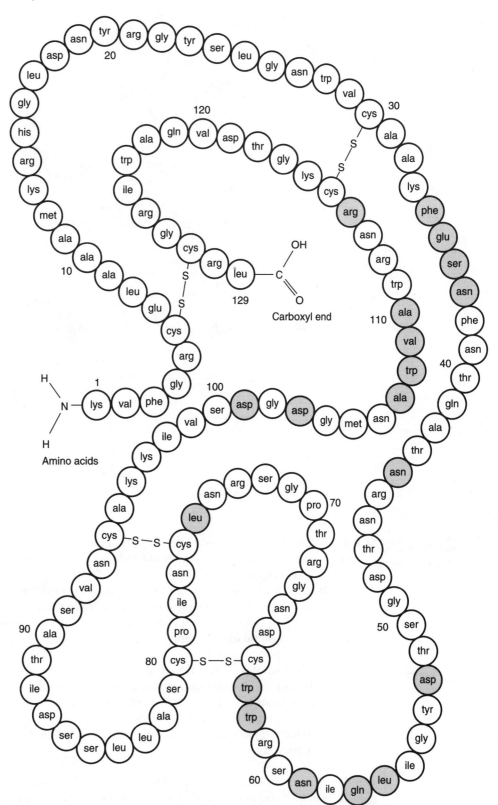

Figure 28
Secondary and tertiary
structure of lysozyme.

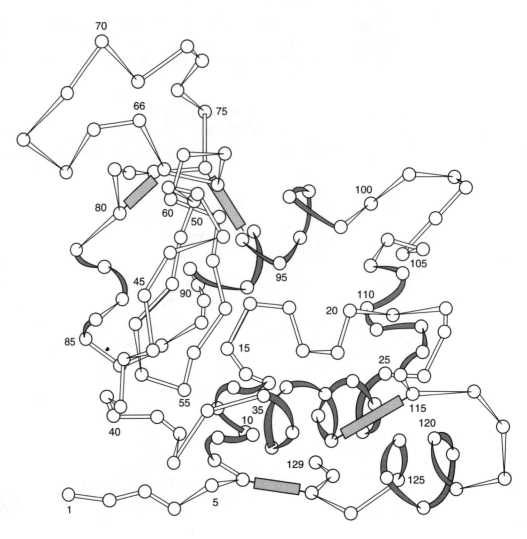

EXERCISE 64

Answers on page 188

a On Fig. 28, trace along the outline of the protein chain starting at the amino end, residue 1. List the amino acid residue numbers of the areas which show α-helical secondary structure.

b The chain is held together in four places by disulphide bridges.
   i)   State which amino acid forms disulphide bridges.
   ii) Write an equation for the reaction in which a disulphide bridge is formed. (Assume that the amino and carboxyl groups of each amino acid are involved in peptide bonds.)
   iii) What type of reaction is this?

c In addition to hydrogen bonding, another type of bonding can operate between certain amino acid residues. Write an equation to show the formation of such a bond between glutamic acid and lysine residues.

d What happens to the structure of lysozyme when it is 'denatured' and how may this be achieved?

We end this section of work on protein structure with a very short experiment which enables you to detect the presence of a protein in an aqueous mixture.

EXPERIMENT 6 Biuret test for proteins

Aim   This experiment is intended to give you practical experience of the biuret test for proteins.

Introduction   The biuret test is based on a coloured complex formed between copper(II) ions and the peptide links of neighbouring protein chains. You carry out the test on examples of available protein material.

Requirements   ■ safety spectacles
■ 4 test-tubes in a rack
■ spatula
■ protein sample(s) (e.g. egg albumin, gelatin, fresh milk)
■ wash-bottle of distilled water
■ sodium hydroxide solution, 2 M NaOH
■ teat-pipette
■ copper(II) sulphate solution, 0.1 M CuSO$_4$

HAZARD WARNING

Sodium hydroxide is corrosive. Therefore you **must**:
■ **wear safety spectacles.**

Procedure

1. Take a small amount (enough to cover a spatula tip if solid, about 2 cm^3 in the bottom of a test-tube, if liquid) of one of the proteins. Dissolve it in water so that the total volume is no more than a quarter of a test-tube. If necessary, warm the tube.
2. Allow the contents of the tube to cool.
3. Add an equal volume of 2 M sodium hydroxide solution followed by five drops of 0.1 M copper(II) sulphate solution.
4. Leave the tube to stand, if necessary, and note the colour that develops.
5. Repeat steps 3 and 4 using water instead of protein solution. Compare the colour of this 'blank' tube with the colour of your protein solution from 3. If you don't get a definite colour, try again with a more concentrated protein solution in step 1.
6. Repeat steps 1 to 4 using other protein(s).

Questions
Answers on page 189

1. How do you think the biuret test might be used to estimate the concentration of a protein in a solution?
2. The biuret test is a test for all compounds containing two peptide links at a suitable separation for use as a bidentate ligand. Given the fact that the Cu^{2+} ion forms 4-coordinate, planar complexes, suggest a possible formula for the complex between Cu^{2+} and portions of two protein chains with the general formula:

$$\begin{array}{cccccc} & H & H & & H & H \\ & | & | & & | & | \\ -C & -N & -C & -C & -N & -C- \\ \| & & | & \| & & | \\ O & & R & O & & R \end{array}$$

3. If you hydrolysed a protein sample and converted it completely into its constituent amino acids, would you expect to get a positive biuret test? Explain.

We referred to X-ray crystallography at several points in the last section as a method of structure determination. In the next section we look at the chemical methods, amino acid analysis and sequence determination, which are used to complement X-ray analysis. Check with your teacher whether this is a requirement of your particular syllabus.

■ 5.6 Determining the primary structure of a protein

It is relatively easy to hydrolyse a protein completely into separate amino acids which can be identified by chromatography; you learn about the technique in Experiment 7. However, working out the order in which the amino acids are joined together is much more difficult – a kind of chemical jigsaw puzzle.

A technique for the sequence determination of amino acids in a protein chain was worked out over several years by Sanger in the 1920s. Additional methods have now allowed the processes to be automated so that what once took months (even years) can now be done in hours.

OBJECTIVES

When you have finished this section you should be able to:
■ state the reaction of **1-fluoro-2,4-dinitrobenzene (FDNB)** with an amino acid;
■ explain what is meant by the **'N-terminal'** and **'C-terminal'** ends of a polypeptide chain;
■ explain the role of **enzymes** such as trypsin and carboxypeptidase in **sequence determination**;
■ explain how **N-terminal analysis** of a polypeptide is carried out.

Read about sequence determination in a polypeptide. You may find this listed in the index under **Sanger**. Look for an account of the way in which partial hydrolysis of a protein can be controlled. For a reasonably full description you may have to use an organic chemistry textbook (rather than a general one) or a simple biochemistry book.

One of Sanger's techniques concerned the reaction between 1-fluoro-2,4-dinitrobenzene and an amino group. You explore this reaction in the next exercise.

EXERCISE 65

Answers on page 189

a Write the formula for 1-fluoro-2,4-dinitrobenzene.

b Give an equation to show how FDNB reacts with alanine, $CH_3CH(NH_2)CO_2H$.

c The bond between FDNB and an amino acid is resistant to hydrolysis. In what way would this be useful in the sequence determination of a polypeptide?

Another technique used in sequence determination is to digest a peptide with carboxypeptidase for about 12 hours. This enzyme releases amino acids, sequentially, from the C-terminal end of the peptide. You have probably met the technique in your reading. We include an example in the following exercises.

EXERCISE 66

Answers on page 190

A pentapeptide was isolated from a certain bacterium. Several tests were carried out on the pentapeptide and their results are given below. Use the information to suggest a sequence for the amino acids in the peptide.

A. On total hydrolysis, the following amino acids were identified, in the given proportions:

arginine (1), glycine (2), lysine (1), valine (1).

B. On partial hydrolysis, the following dipeptides were identified: gly–gly; val–gly; lys–val; gly–arg.

C. 1.00 g of the pentapeptide, whose relative molecular mass is 501, was treated with carboxypeptidase for twelve hours. The amino acids released were:

arginine: 2.1×10^{-3} mol   glycine: 4.3×10^{-3} mol

D. The pentapeptide was treated with 1-fluoro-2,4-dinitrobenzene (FDNB). After hydrolysis DNP-lysine was identified by chromatography.

EXERCISE 67

Answer on page 190

The amino acid sequence in a peptide can, in favourable circumstances, be determined from the amino acids formed on hydrolysis and the dipeptides formed on partial hydrolysis. A certain tetrapeptide on hydrolysis gave alanine and glycine in the molecular ratio of 3 : 1. Describe how such a result can be established. On partial hydrolysis, the only dipeptides formed were glycylalanine and alanylalanine. Describe how the structures of these dipeptides can be established experimentally. Suggest a structure for the tetrapeptide.

We now examine the technique of paper chromatography, which has played a vital part in the determination of primary structure.

■ 5.7 Chromatography

All chromatographic methods depend on similar principles. In this section we concentrate on paper chromatography, but you may also need to be able to outline other techniques such as thin-layer chromatography (TLC), column chromatography and gas–liquid chromatography (GLC). Ask your teacher how much you should study.

When you have finished this section you should be able to:

■ outline the role of **chromatography** in the determination of the **primary structure** of a protein;

■ explain the terms **stationary and mobile phase**, **eluent and adsorbant**, **adsorption and partition**, in the context of chromatography;

■ outline the distinguishing features of four types of chromatography – column, thin-layer, paper and gas–liquid;

■ explain the difference between **ascending** and **descending** chromatography;

■ explain the use of **ninhydrin** solution in detecting amino acids on chromatograms;

■ explain what an R_f **value** is and calculate such values for substances on a paper chromatogram;

■ carry out a simple **separation of amino acids** using paper chromatography.

Find a textbook that has a section on chromatography. As the technique is more important in organic chemistry, you should find a more detailed treatment in a separate organic textbook. Scan the whole section, including column, thin-layer, paper and gas–liquid chromatography. You need to identify the principle behind each method as well as the substances used in it and the main factors affecting the process.

You should now do the next exercise, which is to help you summarise and record what you have learned from your reading.

EXERCISE 68

Answers on page 190

Table 14

a Complete a copy of the following table to show the characteristics of the four main types of chromatography.

Type of chromatography	Separation phases		Principle* – adsorption or partition
	Mobile	**Stationary**	
Column	Liquid	Solid	Adsorption[†]
Thin-layer			
Paper			
Gas–liquid			

* In some cases, **both** adsorption and partition occur, but one is nearly always predominant.
[†] Partition between eluent and adsorbed liquid may be important in some cases.

b Which substances are commonly used on the stationary phase in column and thin-layer chromatography?

c In column chromatography, the substances making up the stationary phase are usually used in an 'activated' form.

i) Explain what happens when these substances are activated.

ii) How does the activated form affect the progress of a substance in a chromatography column?

d What are the main factors affecting the rate at which substances separate?

We do not discuss gas–liquid chromatography further in this book. However, if it is available, you may wish to see part of the ILPAC video programme, 'Instrumental Techniques', which illustrates the procedure.

Now you can gain some practical experience of chromatography. We have chosen a method using paper because it lends itself to a simple and quick experiment although, as you have learned from your reading, the other types of chromatography are now of more importance. However, you can apply your knowledge of paper chromatography to the other techniques if you need to study them as well.

If it is available, watch the ILPAC video programme 'Organic techniques III', in which we demonstrate paper chromatography and explain the calculation of R_f values.

If you are unable to watch the video programme, read about R_f values in a suitable textbook and make sure you know how to calculate them.

EXPERIMENT 7 Paper chromatography

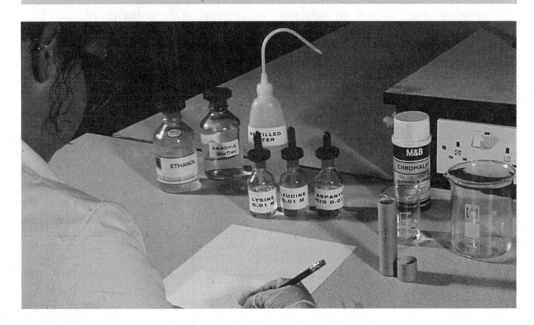

Aim
The purpose of this experiment is to illustrate the use of paper chromatography for the separation and identification of amino acids.

Introduction
In this experiment you separate a mixture of three amino acids by means of paper chromatography. From the chromatogram you calculate R_f values for the individual amino acids.

Requirements
- safety spectacles and protective gloves
- measuring cylinder, 10 cm^3
- beaker, 400 cm^3, tall-form
- watch glass (big enough to cover beaker)
- ethanol, C_2H_5OH
- wash-bottle of distilled water
- ammonia solution, 0.880 NH$_3$
- square of chromatography paper, 12.5 cm × 12.5 cm
- pencil and ruler
- sheet of file paper
- 4 melting-point tubes
- aspartic acid solution, 0.01 M
- leucine solution, 0.01 M
- lysine solution, 0.01 M
- mixture of the three amino acids above
- 2 retort stands, bosses and clamps
- 2 paper clips
- hair dryer
- ninhydrin aerosol spray
- oven (105 °C)

Procedure

1. In a fume cupboard, prepare the solvent mixture by pouring the following into a 400 cm³ tall-form beaker:

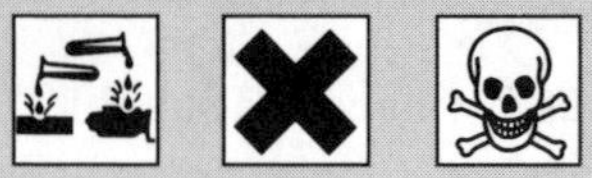

   24 cm³ of ethanol
   3 cm³ of distilled water
   3 cm³ of 0.880 ammonia

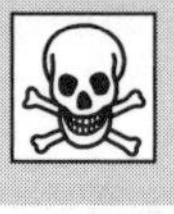

2. Cover the beaker with a watch glass. Swirl to mix the liquids and leave to stand.
3. Handling it only by the top edge, place a square of chromatography paper on a clean sheet of file paper. With a pencil (**not** a pen) draw lines and labels as shown in Fig. 29.

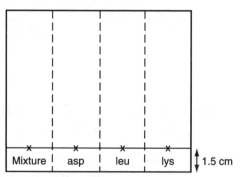

4. Dip a clean melting point tube into the solution of mixed amino acids and then touch it **briefly** on the appropriate labelled cross so that a spot, no more than 5 mm across, appears on the paper.
5. Using a fresh tube each time, repeat step 4 for each of the three solutions of single amino acids.
6. Place a **clean** ruler with its edge along one of the dashed lines and hold it firmly in place with one hand. Without touching it with your fingers, fold the chromatography paper along the line by sliding your hand under the **file paper** and lifting.
7. Repeat the folding procedure for the other two lines so that the opposite edges of the paper almost meet to form a square cross-section.

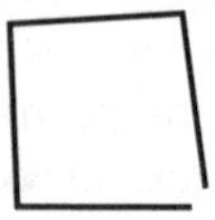

8. Hold the paper by the edge furthest from the start line, and place it in the beaker so that it does not touch the sides. Replace the cover and leave to stand.
9. Clamp the ruler horizontally at a height of 20–30 cm between two retort stands in a fume-cupboard. This is to support the chromatography paper for drying when the run has finished.
10. While you are waiting, get on with some other work, but look at the paper every ten minutes to see how far the solvent has soaked up the paper.

11. When the solvent has reached nearly to the top of the paper (30–40 minutes) or when you have only 15 minutes laboratory time left, whichever is the sooner, remove the paper from the beaker, open it out and clip it on to the ruler to dry. You can hasten the drying with a hair dryer.
12. When the paper is dry, spray it evenly with ninhydrin solution. Dry it again and then heat it in an oven at 105 °C for five minutes.
13. Remove the paper from the oven and mark with a pencil the positions of the coloured spots.
14. Measure the distances from the origin line to the centres of the spots and record them in a copy of Results Table 5.

Results and calculations

Calculate an R_f value for each spot as follows:

$$R_f = \frac{\text{distance travelled by spot}}{\text{distance travelled by solvent}}$$

Results Table 5

Amino acid	Distances travelled/cm		
	By solvent	By amino acid	R_f value
Aspartic acid – alone – in mixture			
Leucine – alone – in mixture			
Lysine – alone – in mixture			

Specimen results on page 191

Questions

Answers on page 191

1. For each amino acid, compare your two R_f values with each other and with our specimen results. Why do you think there is some variation?
2. Why do R_f values change when a different solvent is used?
3. Why is it so important to avoid touching the chromatography paper with your fingers?

A protein hydrolysate may contain as many as twenty different amino acids and it is not possible to separate all of them completely by the simple method you have just used. Whichever solvent is chosen, there is always at least one pair of amino acids with almost identical R_f values.

To overcome this problem, two-way paper chromatography was developed. A single spot of mixture is placed in one corner of the paper and is partially separated by one solvent in the usual way into a line of spots, some of which contain more than one amino acid. The paper is then dried, turned through 90° and placed in a solvent of different character from the first. The solvents are chosen so that no two amino acids have the same R_f values in both; this means that a complete separation can be done on a complex mixture.

It is notoriously difficult to obtain reproducible results using paper chromatography – R_f values can vary widely from one experiment to another. For this reason, two-way chromatography is more often done on glass plates covered with a layer of solid adsorbant, but the general appearance of the chromatograms, and the calculation of R_f values, are much the same.

The next exercise tests your understanding of two-way chromatography.

EXERCISE 69

Answers on page 191

Table 15

Table 15 gives data about a number of amino acids that occur in proteins.

Name and abbreviation		Relative molecular mass	R_f value in phenol	R_f value in butan-1-ol/ ethanoic acid
Alanine	ala	89	0.43	0.38
Aspartic acid	asp	133	0.13	0.24
Glycine	gly	75	0.33	0.26
Leucine	leu	131	0.66	0.73
Lysine	lys	146	0.62	0.14
Phenylalanine	phe	165	0.64	0.68
Serine	ser	105	0.30	0.27
Valine	val	117	0.58	0.40

A small polypeptide was hydrolysed with 6 M acid and the resulting amino acids were separated by two-way chromatography. The chromatogram is reproduced below.

Figure 30

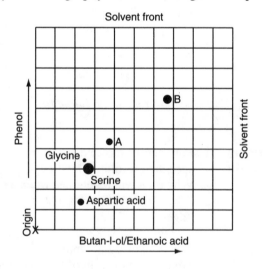

a Determine the R_f values in both solvents of the amino acids labelled **A** and **B** on the chromatogram, and identify them.

b The polypeptide was reacted with 1-fluoro-2,4-dinitrobenzene (FDNB). After hydrolysis, DNP-glycine was identified by chromatography. What information does this give about the sequence of amino acids in the polypeptide?

c When 1.00 g of the polypeptide, whose molar mass was found to be about 780 g mol^{-1}, was subjected to the action of carboxypeptidase for 12 hours, the amino acids released were:

serine	2.5×10^{-3} mol
aspartic acid	1.2×10^{-3} mol
amino acid A	1.1×10^{-3} mol

What information does this result give about the sequence of amino acids in the polypeptide?

d Quantitative estimation of the amino acids in the polypeptide showed that they were present in the following molar ratio:

serine	3
aspartic acid	2
amino acid A	1
amino acid B	1
glycine	1

Using this information and information from the previous sections, which of the following is the most probable sequence of amino acids in the polypeptide?
i) (NH_2)–gly–ser–B–asp–ser–asp–A–(CO_2H)
ii) (NH_2)–gly–asp–A–ser–B–asp–ser–ser–(CO_2H)
iii) (NH_2)–gly–ser–B–asp–A–asp–ser–ser–(CO_2H)
iv) (NH_2)–ser–ser–asp–A–ser–B–asp–gly–(CO_2H)
v) (NH_2)–A–asp–ser–ser–ser–B–asp–gly–(CO_2H)
Explain your reasoning for each of the sequences (i) to (v).

The next exercise takes the form of five multiple-choice questions based on a practical situation.

EXERCISE 70
Answers on page 192

Questions 1–5 refer to the experiment described below. Select the best answer for each question.
In order to determine which amino acid in a mixed dipeptide has a free amino group, these procedures are adopted.
i)   The dipeptide, dissolved in sodium hydrogencarbonate solution, is warmed with excess 1-fluoro-2,4-dinitrobenzene dissolved in ethanol, to give a dinitrophenyl peptide.
ii) The mixture is then extracted with ether.
iii) The aqueous layer is heated with a little concentrated hydrochloric acid.
iv) A second ether extraction is carried out, the ether removed and the residue from the ether extract examined chromatographically.

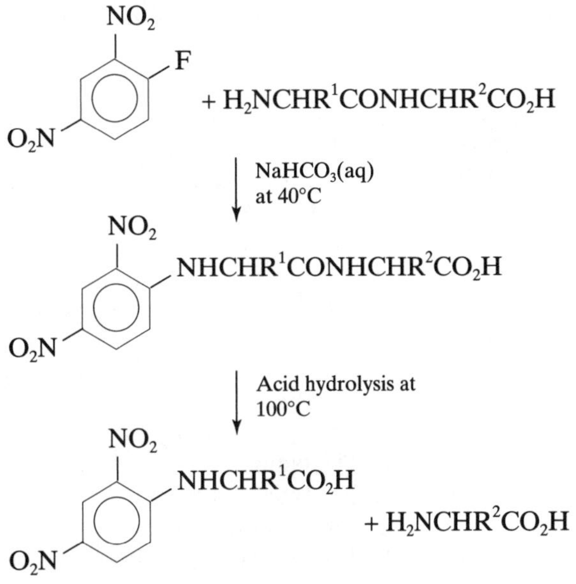

NO_2

F

O_2N

$+ H_2NCHR^1CONHCHR^2CO_2H$

$NaHCO_3(aq)$ at 40°C

NO_2

$NHCHR^1CONHCHR^2CO_2H$

O_2N

Acid hydrolysis at 100°C

NO_2

$NHCHR^1CO_2H$

O_2N

$+ H_2NCHR^2CO_2H$

1. The purpose of the first ether extraction is to remove:
   A fluoride ions,
   B sodium hydrogencarbonate,
   C alcohol,
   D unchanged dipeptide,
   E unchanged 1-fluoro-2,4-dinitrobenzene.

2. The second amino acid ($H_2NCHR^2CO_2H$) from the dipeptide:
   A dissolves in the ether in the first ether extraction,
   B dissolves in the ether in the second ether extraction,
   C is destroyed in the acid hydrolysis,
   D remains in the aqueous layer after the second ether extraction,
   E forms a fluoro-derivative with the fluoride ions.

3. A number of yellow spots appeared on the chromatogram. Which one corresponds to a dinitrophenyl amino acid with an R_f value of 0.30–0.35?

Figure 31

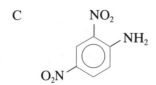

Original spot placed here

4. A student found four other yellow spots on the chromatogram in addition to the dinitrophenyl amino acid. Which of the following compounds would be **least** likely to be present?

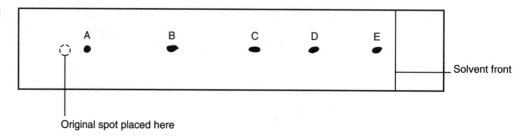

5. Which of the following would have the greatest effect on the R_f value of the dinitrophenyl amino acid?
   A The type of paper used in the chromatogram.
   B The temperature at which the chromatogram is run.
   C The type of solvent used to develop the chromatogram.
   D The quantity of substance used in the original spot.
   E The direction in which the solvent is run, e.g. upwards or downwards.

A further development in the separation and identification of amino acids is the use of ion-exchange resins in column chromatography, using high pressure to speed up the elution. This technique has been automated so that separations can be done very efficiently and rapidly. You may wish to read about ion-exchange chromatography in an up-to-date textbook.

You may now consolidate your knowledge of amino acids and proteins by attempting the following teacher-marked exercise.

EXERCISE
Teacher-marked

Amino acids join together via peptide bonds to form proteins. The structures of two naturally occurring amino acids are shown below.

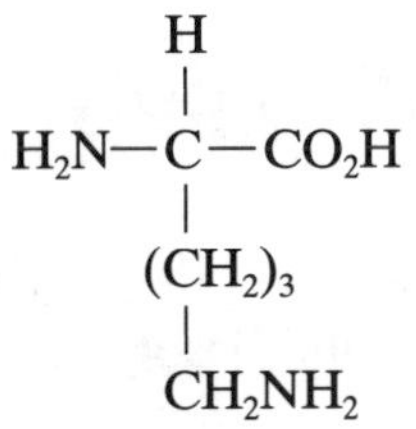

$$H_2N-\underset{\underset{CH_2NH_2}{\overset{|}{(CH_2)_3}}}{\overset{\overset{H}{|}}{\underset{|}{C}}}-CO_2H \qquad H_2N-\underset{\underset{H}{|}}{\overset{\overset{H}{|}}{C}}-CO_2H$$

lysine 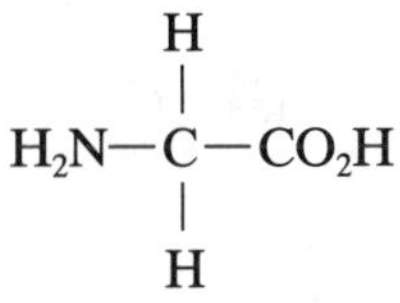   glycine

a Draw the structure of two dipeptides which can be formed by condensing lysine and glycine. Clearly indicate the peptide linkages, showing all bonds.

b Describe, and state the bonds present in, the secondary structure of a protein.

c State the two types of bond responsible for the maintenance of the tertiary structure of a protein.

d Amino acids exist as zwitterions in aqueous solution. Draw the structural formula of the zwitterion formed from lysine, and write equations to show how it can act as a buffer.

e Only one of the amino acids glycine or lysine is optically active.
   i)   Identify the optically active amino acid and the feature that led you to this conclusion.
   ii) Draw diagrams to illustrate the two optical isomers of this amino acid.

Another important group of biological polymers, nucleic acids (DNA), is covered in Appendix 2 – check whether this is a particular requirement of your syllabus.

Having completed a study of natural polymers, we now consider synthetic ones – plastics.

6 SYNTHETIC POLYMERS

Having looked at natural polymers, i.e. polysaccharides in ILPAC 8, Functional Groups, and proteins in the previous chapter, we now turn our attention to the synthetic polymers – plastics. You were introduced to addition polymerisation in ILPAC 5, Introduction to Organic Chemistry. We refresh your memory on this in a preliminary exercise and then go on to study the other type – condensation polymerisation. Since syllabus requirements vary on this particular topic you must check with your teacher or syllabus to find out which parts apply to you. Some of the material will be more relevant for those studying a special option or module on polymers.

■ 6.1 Addition polymers – a revision exercise

OBJECTIVES
When you have finished this section you should be able to:
■ show how **addition polymers** are related to ethene, and write equations for the formation of some examples;
■ distinguish between **atactic** and **isotactic** polymers.

The most common type of addition polymer is based on ethene. Refresh your memory on addition polymerisation from ILPAC 5, Introduction to Organic Chemistry (Section 3.6), and then attempt the following exercises.

EXERCISE 71
Answers on page 192

Figure 32 shows how different side groups (Y) on ethene give rise to a variety of different polymers. You complete a copy of it by identifying the side group Y and naming the monomer and polymer in each case. One has already been done for you.

Figure 32
Ethene-based addition polymers.

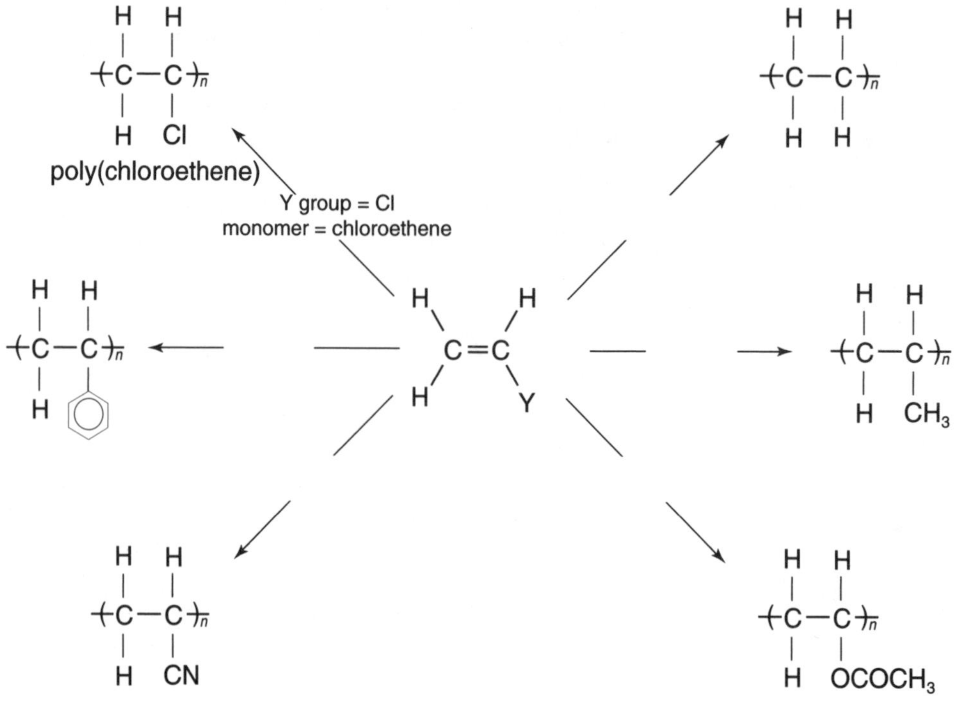

$$+\!\!\begin{array}{c} H \\ | \\ C \\ | \\ H \end{array}\!\!-\!\!\begin{array}{c} H \\ | \\ C \\ | \\ Cl \end{array}\!\!+_n$$

poly(chloroethene)

Y group = Cl
monomer = chloroethene

EXERCISE 72

Answer on page 192

The setting of superglue is an example of addition polymerisation. Superglue is simply a very pure sample of the monomer methyl-2-cyanopropenoate. The vapour of this compound is poisonous and the glue has been known to stick fingers together. Very small amounts of water or bases on the surface of the object are sufficient to initiate the polymerisation and bind articles together. Complete the following equation to show the structural formula of the polymer of superglue, showing at least two repeat units.

$$CH_2= C \begin{array}{c} CO_2CH_3 \\ | \\ | \\ CN \end{array} \longrightarrow$$

methyl-2-cyanopropenoate

Many geometrical arrangements of the polymers shown in Fig. 32 may occur. If you imagine the main carbon chain to be in a horizontal plane then the substituent Y may either stick up above or below the plane. This will, of course, give rise to isomeric polymers with different physical properties. You identify two types in the next section.

■ 6.2 Isotactic and atactic polymers

In most polymers the monomer units always link together so that the substituent groups, e.g. Cl in poly(chloroethene), appear regularly on alternate carbon atoms. However, this type of addition does not necessarily give such an ordered structure as you might think at first by looking at 'flat' formulae with straight chains.

Bonds occur in tetrahedral directions around each carbon atom, and alternate carbon atoms are asymmetric (attached to four different atoms or groups). This means that two different orientations are possible for each monomer unit added. Consequently, there is a huge number of stereoisomers but, fortunately, you need consider two types only.

Read about the structural difference between isotactic and atactic polymers so that you can do the next exercise.

EXERCISE 73
Answers on page 192

Structures **A** and **B** represent portions of polymer chains. Classify each as atactic or isotactic.

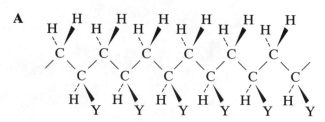

As you might expect, isotactic and atactic versions of the same polymer show differences in some physical properties. We return to this point later. First we consider some polymers formed by condensation reactions between monomer molecules.

■ 6.3 Condensation polymers

OBJECTIVES

When you have finished this section you should be able to:
■ quote some examples of **condensation polymers** and write equations to show their formation;
■ state the difference between a **homopolymer** and a **copolymer**.

Start by reading about the condensation polymers, polyamides and polyesters in a textbook, looking out for the difference between copolymers and homopolymers, and for the types of reaction by which they are formed. You may also wish to revise work from earlier chapters where you studied the condensation reactions between — OH and — CO₂H and between — NH₂ and — CO₂H in amino acids. Don't forget to check with your teacher or syllabus on which parts apply to you.

First, we look at the formation of polyamides.

■ 6.4 Polyamides

Nylons are **polyamides**. Different types of nylon with slightly different properties are formed depending on the lengths of the carbon chains of the monomer(s). You meet three of these in the next three exercises.

EXERCISE 74
Answers on page 192
Figure 33

Figure 33 shows a molecule containing both an amino group and a carboxylic acid group. The box represents a carbon skeleton separating the two groups.

$$H_2N—\boxed{}—CO_2H$$

a Write an equation to show how three such molecules might join together in a condensation reaction. (Don't worry about reaction conditions at this stage.)
b What name is given to the structural unit linking the boxes?

The reaction you have just outlined is the basis of one form of nylon, nylon-11. This uses the 11 carbon molecule $NH_2(CH_2)_{10}CO_2H$, 11-amino undecanoic acid (the name nylon-11 refers to the 11 carbon atoms). In the next exercise you examine a reaction that gives another form of nylon.

Nylon fibres during manufacture.

EXERCISE 75

Answers on page 193

Figure 34

Figure 34 shows two molecules which together can form a condensation polymer.

$$H_2N—\boxed{}—NH_2 \qquad HO_2C—\boxed{}—CO_2H$$

a Write an equation to show how two such molecules might join together.
b If the reaction continued, a polymer chain called a polyamide would be produced. Write down the formula of the repeating unit.
c Explain the meaning of the terms 'copolymer' and 'homopolymer', using the polymer you have just written about and the one in the previous exercise as illustrations.

When each of the molecules in the previous exercise contain six carbon atoms, polymerisation leads to nylon-66:

$$n NH_2(CH_2)_6NH_2 + n CO_2H(CH_2)_4CO_2H \rightarrow \left(\begin{array}{c} H \\ | \\ N—(CH_2)_6—N—C—(CH_2)_4—C \\ \quad\quad\quad | \quad\quad || \\ \quad\quad\quad H \quad\quad O \end{array} \begin{array}{c} O \\ || \\ \\ \end{array} \right)_n + 2n H_2O$$

hexane-1,6-diamine   hexane-1,6-dioic acid

Repeating unit of nylon-66

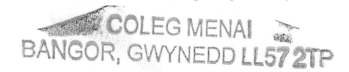

A third method of producing nylon involves a substance called caprolactam. This can be regarded as an internal amide of 6-aminohexanoic acid:

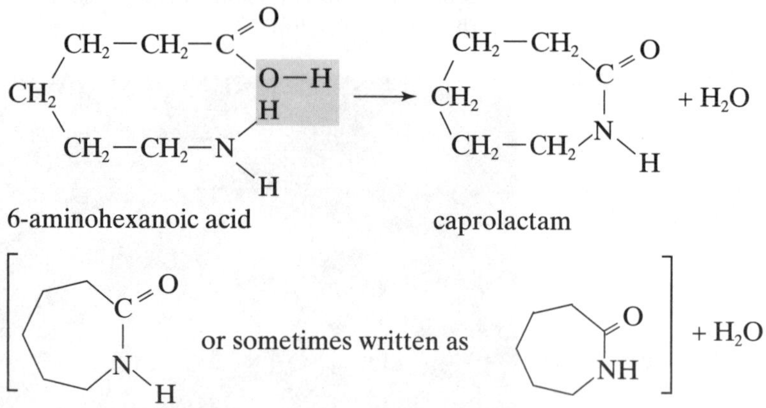

6-aminohexanoic acid   caprolactam

or sometimes written as $+ H_2O$

Like 1,6-diaminohexane and 1,6-hexanedioic acid, caprolactam is made commercially from cyclohexane.

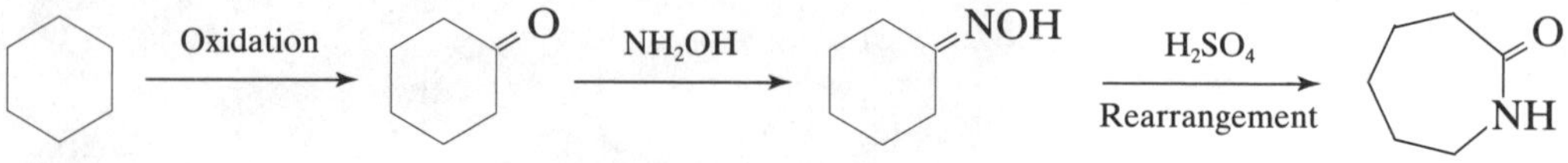

During polymerisation the rings re-open and then link together in long chains.

EXERCISE 76
Answers on page 193

a Write the formula of the repeating unit in the homopolymer made from caprolactam.
b What is the name given to this type of nylon?
c Compare the structure of proteins with nylon. In what ways are they:
  i) similar,
  ii) different?

Now test your understanding of polyamide formation by attempting the following exercises.

EXERCISE 77
Answers on page 193

Choose, from the following list of monomers, those that could polymerise to give each of the structures **A**, **B** and **C**. Which are copolymers and which are homopolymers?

1. $NH_2CH_2CH_2NH_2$

2. $NH_2CH_2CH_2CO_2H$

3. $CO_2HCH_2CH_2CO_2H$

4. $NH_2CH_2CHCO_2H$
            |
            CH_3

5. $NH_2CH_2CHCH_2CO_2H$
              |
              CH_3

A

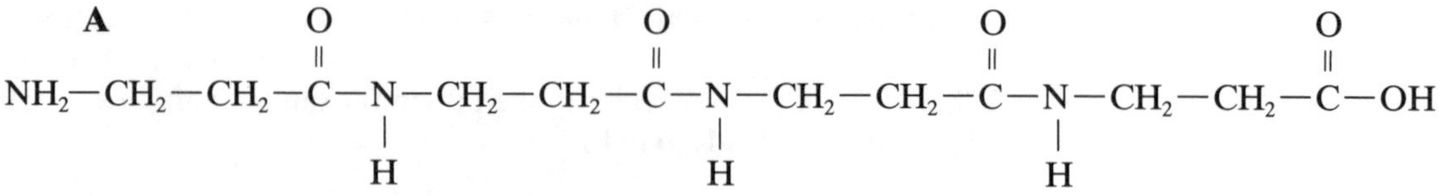

B

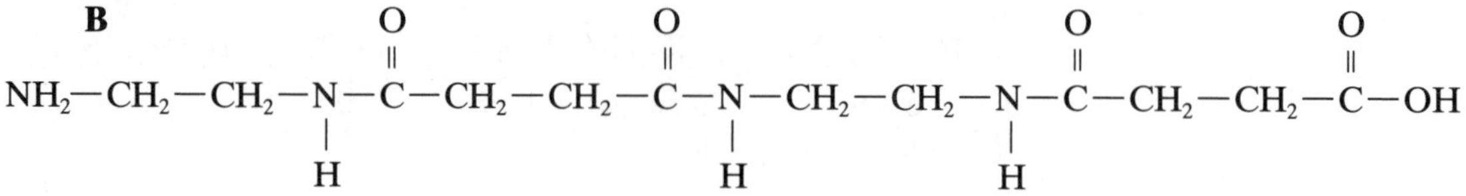

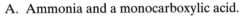

$$\underset{CH_3}{NH_2-CH_2-\underset{|}{CH}-\overset{\overset{O}{\|}}{C}-\underset{H}{N}-CH_2-\underset{CH_3}{\underset{|}{CH}}-\overset{\overset{O}{\|}}{C}-\underset{H}{N}-CH_2-\underset{CH_3}{\underset{|}{CH}}-\overset{\overset{O}{\|}}{C}-\underset{H}{N}-CH_2-\underset{CH_3}{\underset{|}{CH}}-\overset{\overset{O}{\|}}{C}-OH}$$

EXERCISE 78

Answer on page 193

Which one of the following pairs of compounds might be made to combine together, under suitable conditions, to form a polyamide?

A. Ammonia and a monocarboxylic acid.
B. A mono-amine and a monocarboxylic acid.
C. A diamine and a monocarboxylic acid.
D. A mono-amine and a dicarboxylic acid.
E. A dicarboxylic acid and a diamine.

We now consider another type of condensation polymer – **polyesters**. Start by reading the section on polyesters in your textbook to help you do the exercises in the next section.

■ 6.5 Polyesters

Most sails are now made of polyester.

Polyesters are made by the reaction of a diol with a dicarboxylic acid. This is another example of condensation polymerisation, as you demonstrate in the next exercise.

EXERCISE 79
Answers on page 193

a Write an equation to show how a trimer could be formed from the following two molecules:

HO$_2$C——⟨◯⟩——CO$_2$H   2CH$_2$OH—CH$_2$OH

benzene 1,4-dicarboxylic   ethane-1,2-diol
acid (terephthalic acid)   (ethylene glycol)

b What type of linkage is formed between the molecules?
c Which well-known polyester is formed by a similar reaction?

The polyester mentioned in your answer to (c) is melted and forced through fine holes to make a very useful fibre (see Fig. 40 on page 98).

EXERCISE 80
Answer on page 193

Which of the following would react together, in pairs, to form a polymer?
1. HOCH$_2$CH$_2$OH and HO$_2$CCH$_2$CH$_2$CO$_2$H
2. HO$_2$CCH$_2$CH$_2$CO$_2$H and HO(CH$_2$)$_4$CO$_2$H
3. HO(CH$_2$)$_6$OH and ClOC(CH$_2$)$_4$COCl
4. HO$_2$C(CH$_2$)$_4$OH and HOCH$_2$CH$_2$OH

To help consolidate your knowledge about the two main types of polymer, attempt the following teacher-marked exercise. Look through your notes and make a rough plan before you start writing.

EXERCISE
Teacher-marked

Explain what you understand by the terms addition polymerisation and condensation polymerisation, illustrating each answer with an example.

All the polymers you have considered so far have one thing in common – each consists mainly of unbranched chains. We now take a look at polymers formed from monomers with **three** (or more) reactive sites per molecule. These can form multi-dimensional polymers with branched or cross-linked chains.

■ 6.6 Bakelite – a cross-linked polymer

We have already mentioned Bakelite in ILPAC 8, Functional Groups, page 82 when discussing the uses of phenol and methanal, from which it is made. It was discovered in the 1870s (probably the first modern 'plastic') and has been widely used ever since 1907 when Baekeland took out the first patents in New York. Other polymers with superior properties have been developed but Bakelite is cheap and is used as an insulator of both heat and electricity in applications where its brittleness and dark colour are not serious drawbacks.

OBJECTIVES

When you have finished this section you should be able to:
■ give an example of a **cross-linked polymer**;
■ describe the effect of cross-linking on the physical properties of a polymer.

Start by reading about **Bakelite** looking, in particular, for an explanation of its insolubility and brittle nature. You may find it called **phenol-methanal polymer**, or **PF (phenol-formaldehyde) resin**.

In the first stage of the reaction between phenol and methanal, phenol is substituted at the 2,4 and 6 positions of the ring. In the next exercise you suggest which products are formed.

EXERCISE 81
Answers on page 193

a Write an equation to show how one molecule of methanal could react with one molecule of phenol.
b Do you think this reaction proceeds by nucleophilic or electrophilic attack? Explain.
c Write the formulae of products that would be formed in further reactions if there were plenty of methanal present in the reaction mixture.

The reaction you have just described results in mono-, di- and tri-substituted phenol molecules, depending on the relative amounts of phenol and methanal present. In the next stage of the reaction, the phenol-alcohols condense together (or with more phenol) to form chains, as described in the following exercise.

EXERCISE 82
Answer on page 194

Complete the equation below to show the formation of a phenol-methanal trimer.

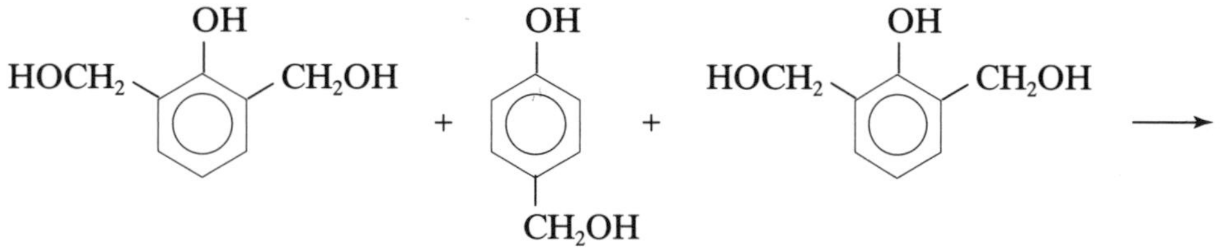

The last stage in the process is extensive cross-linking between long chains. Under appropriate conditions this continues until each reactive site on each phenol molecule is joined via a methylene ($-CH_2-$) group to another phenol molecule. A possible arrangement is shown below but, in practice, the structure would be three-dimensional and less regular than this.

Figure 35
Bakelite.

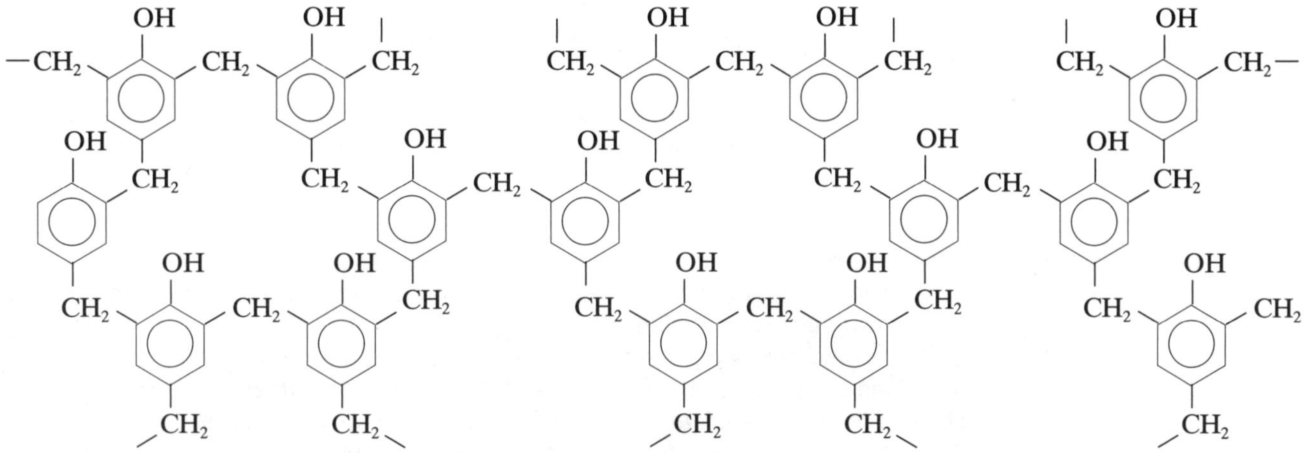

EXERCISE 83
Answer on page 194

Why do you think Bakelite is hard, brittle and insoluble in all known solvents?

You have probably heard of some other cross-linked polymers made from methanal by condensation reactions. If carbamide (urea), NH_2CONH_2, is used in place of phenol, carbamide-methanal resins can be made.

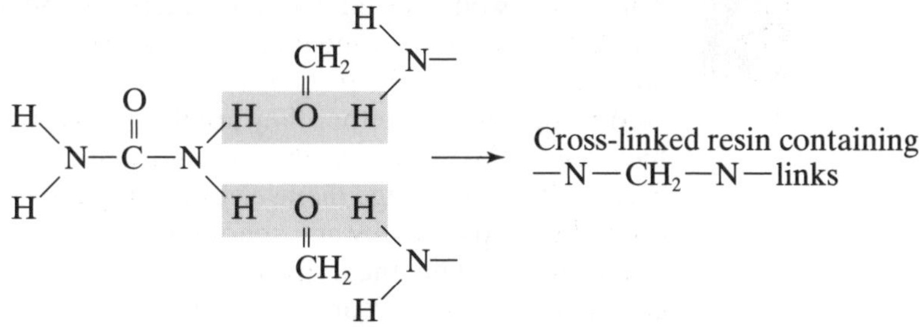

Cross-linked resin containing $-N-CH_2-N-$ links

These polymers are commonly known as urea-formaldehyde (UF) resins and they have been used extensively in foam form for heat insulation, e.g. in cavity walls. However, there have been serious health hazards in some faulty installations because the foam can give off toxic methanal vapour. If you are interested in reading about this, ask your teacher for some references.

Another well-known polymer is made from methanal and 2,4,6-triamino-1,3,5-triazone, commonly called Melamine.

Each molecule has six possible sites for condensation so that very extensive cross-linking occurs. The product is strong and heat-resistant, which makes it suitable for tableware and tough laminates such as Formica.

You have already seen some of the ways in which the physical properties of polymers depend on their structure. In the next section we consider physical properties in more detail.

■ 6.7 Physical properties of polymers

In this section we take an overview of the properties of different polymers, including some of the more important terms used to describe them.

This area is well covered in the polymers section of many textbooks. Read the sections on physical properties and cross-linking. Also find out about the process of vulcanisation and how it changes the properties of rubber.

Most polymers, whether they are thermoplastic or thermosetting, do not appear to have any of the properties that you normally associate with crystals. However, polymer chains often do arrange themselves in a regularly packed manner to give small crystallites separated by amorphous areas. This is shown diagrammatically in Fig. 36.

The more readily the individual chains can be aligned, the greater is the extent of crystalline character. You explore this idea in the next exercise.

Figure 36
Diagram of crystallites in a polymer.

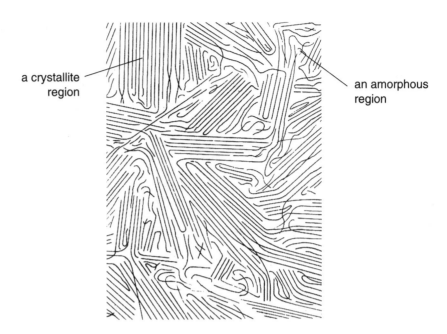

a crystallite region

an amorphous region

EXERCISE 84

Answers on page 194

a What instrumental technique is used to detect the crystalline nature of certain polymers?

b Would you expect the following features to be associated with a high or a low degree of crystalline character in a polymer? Explain.
   i)   A regular 'head-to-tail' arrangement of monomer units.
   ii) An atactic arrangement of the groups attached along the polymer chain.
   iii) Large groups attached along the chain.
   iv) A considerable amount of chain branching.
   v)   Extensive cross-linking between chains.
   vi) Thermosetting properties.

c 'High density' poly(ethene) is made using Ziegler–Natta catalysts which tend to produce long chains with hardly any branching. How does this affect its density and other properties?

In the exercises that follow you consider how the physical properties of different synthetic polymers is related to their structures. You will find it useful to refer to Fig. 37 which shows how synthetic polymers are classified.

Figure 37

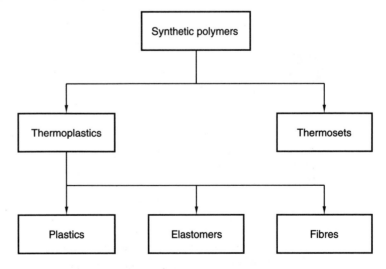

We compare the structures of thermoplastics and thermosets first.

■ 6.8 Thermoplastics and thermosets

EXERCISE 85

Answers on page 194

Figure 38

a Figure 38 shows very simplified representations of the molecules in a thermoset and a thermoplastic polymer. Identify each type with explanations.

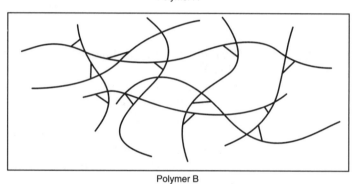

Polymer A

Polymer B

b Phenylethene (styrene), $CH_2 = CHC_6H_5$, polymerises to form poly(phenylethene) (polystyrene) which is a thermoplastic. What is the meaning of the term 'thermoplastic'?
c How do the properties of a thermoset differ from those of a thermoplastic?
d Give an example of a thermosetting plastic.

Bulky side groups, which may or may not be polar, can affect physical properties, as you see in the next exercise.

EXERCISE 86

Answers on page 195

Methyl 2-methylpropenoate can be polymerised to form Perspex:

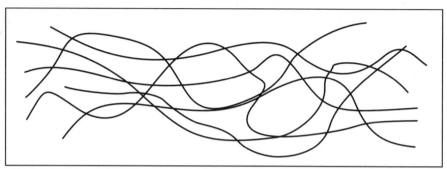

methyl 2-methylpropenoate

a Draw a structural formula for poly(methyl 2-methylpropenoate), Perspex, showing two monomer units.
b Poly(methyl 2-methylpropenoate), Perspex, is a thermoplastic material. However, it has only a limited degree of flexibility; when it is stressed too far, it tends to crack in a brittle manner. Use your knowledge of polar structures to suggest explanations for both these properties.

Extensive cross-linking in three dimensions results in rigid structures like those mentioned above. A smaller degree of cross-linking is found in polymers known as rubbers, which we consider in the next section.

■ 6.9 Elastomers

Polymers which are soft, springy and which can be deformed and then go back to their original shape are called elastomers. You relate this to structure in the next exercise.

EXERCISE 87

Answer on page 195

Figure 39 shows the arrangements of the long-chain molecules of an elastomer unstretched and stretched. Study it and use it to explain why an elastomer will return to its original shape after being stretched.

Figure 39

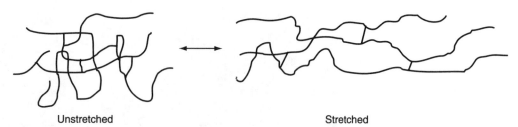

Unstretched   Stretched

We now look at a natural elastomer, rubber, and how the process of vulcanisation changes its properties.

Tapping for rubber.

Pure natural rubber is a long-chain polymer made from the milky latex fluid collected from rubber trees. It is of little practical use because it is not hard-wearing and becomes soft and sticky when warm. It becomes much more useful when the polymer chains are cross-linked by a process called vulcanisation.

EXERCISE 88

Answers on page 195

Natural rubber is a polymer of 2-methylbuta-1,3-diene.
a Write the structural formula of the monomer.
b Show the structure of a section of the polymer.
c How would you classify this polymer?
d What chemical substance is used to vulcanise rubber?
e In what way is the structure modified by vulcanisation?
f What feature of the polymer chain enables vulcanisation to take place?
g How are the properties of natural rubber modified by increasing degrees of vulcanisation?

Rubber tyres.

The uncertain supply of natural rubber, particularly in wartime, led to the development of synthetic rubbers. Some of these have superior properties to natural rubber, but are more expensive. They are related to natural rubber, as indicated below.

$$CH_2 = CH - C = CH_2$$
$$|$$
$$CH_3$$
2-methylbuta-1,3-diene $\longrightarrow$ natural rubber

$$CH_2 = CH - C = CH_2$$
$$|$$
$$Cl$$
2-chlorobuta-1,3-diene $\longrightarrow$ Neoprene

$$CH_2 = CH - CH = CH_2$$
buta-1,3-diene $\longrightarrow$ Buna rubber

$$CH_2 = CH - CH = CH_2 + C_6H_5CH = CH_2 \longrightarrow$$ Buna S rubber (SBR)
buta-1,3-diene   phenylethene   (a copolymer)

The properties of polymers can be altered by other methods. One method is to add a plasticiser, as you now see.

■ 6.10 Plasticisers

Plasticisers are not polymers but consist of longish molecules which get between the polymer chains and reduce the attraction between them, making the chain more flexible. A plasticiser is rather like a lubricant. You explore this idea in the next exercise.

EXERCISE 89
Answers on page 195

Clingfilms.

The compound 'DEMA' is used as a plasticiser in the poly(chloroethene), PVC, film used to make clingfilm for wrapping food. Without a plasticiser the film is too brittle to use. DEMA has been linked to cancer in test animals and since 1987 its use has been decreasing. It was discovered that DEMA migrates from the clingfilm and dissolves in the fat and oil components of the food. It has been replaced as a plasticiser by 'ATBC' but there is now concern that this compound needs to be better understood before it is considered safe.

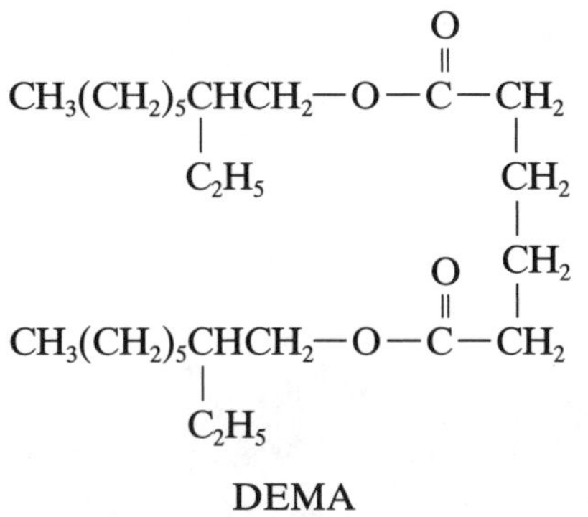

DEMA

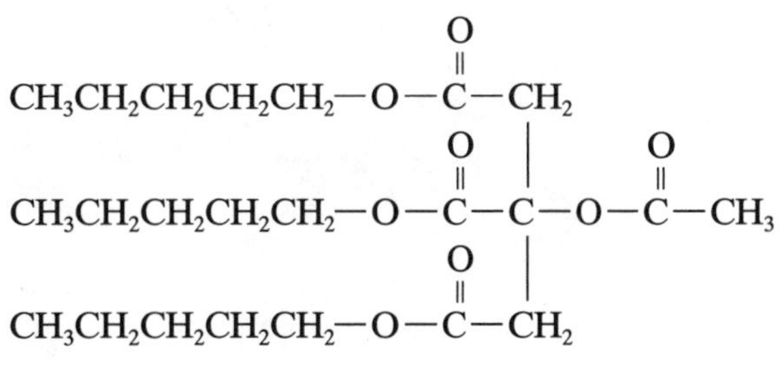

ATBC

a Name the functional group present in both DEMA and ATBC.
b Fats and oils contain the same functional group as is present in plasticisers. Use your knowledge of intermolecular interactions to explain why both the plasticisers will dissolve in fats and oils.

c PVC has the structure:

$$\left(\begin{array}{cc} H & H \\ | & | \\ -C & -C- \\ | & | \\ H & Cl \end{array}\right)_n$$

Explain why PVC is more brittle than poly(ethene).
d Suggest how a plasticiser molecule might reduce the brittle behaviour.
e In an attempt to investigate the structure of ATBC, a chemist hydrolysed some of the substance by refluxing it with dilute hydrochloric acid. Draw full structural formulae of the products you would expect.

Some polymers can be made into fibres as you will see in the next section.

■ 6.11 Fibres

One common method of making polymer fibres is to pass the polymer material through a network of fine holes. The strands of the polymer that emerge are then wound together to give a fibre made up of multiple strands. Figure 40 illustrates this.

Figure 40
One method of making polymer fibres.

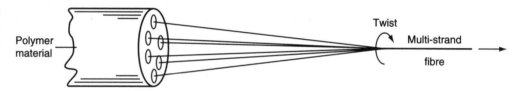

Not all polymers are suitable for making into fibres. One important property is that weak intermolecular forces such as hydrogen bonding or dipole–dipole attraction should take place between chains. You explore this idea in the next exercise.

EXERCISE 90
Answers on page 196

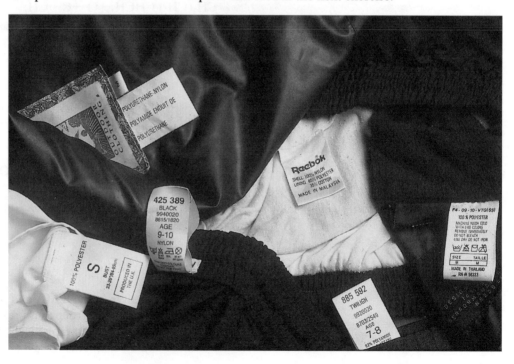

Suggest reasons why propenenitrile,

$$\left(\begin{array}{cc} H & H \\ | & | \\ -C & -C- \\ | & | \\ CN & H \end{array}\right)_n$$

(Orlon), nylon and polyesters are suitable for use in textile fibres whereas poly(phenylethene) and poly(ethene) are not.

In the next section we consider the production of some important polymers.

■ 6.12 Industrial production of polymers

At A-level, you may be required to know the outlines of selected manufacturing processes. (Precise details may be closely guarded secrets belonging to the chemical company concerned.) Before you start this section, check with your teacher which processes you need to know for your particular syllabus.

OBJECTIVES

When you have finished this section you should be able to:
- outline the industrial production of some of the following:
  - **a poly(phenylethene)** (polystyrene),
  - **b poly(ethene)** (polythene), low- and high-density forms,
  - **c poly(chloroethene)** (polyvinylchloride, PVC),
  - **d poly(tetrafluoroethene)** (PTFE),
  - **e nylon,**
  - **f Terylene;**
- state at least one use for each of the polymers listed above.

You should now read about, and make brief notes on, any processes mentioned in your syllabus. As a guide, look for the **conditions** needed: temperature, pressure and details of any catalyst used. Note how many stages there are in the process and whether there are any by-products, including any that are fed back into the main process. You should try to construct simple flow diagrams for processes specified on your syllabus using the generalised flow diagram given in ILPAC 8, Functional Groups, page 40, as a guide. Another important aspect is the **uses** of the polymers – you need to know at least one main use for each polymer you study.

Also look for details of any **hazards** involved in the process, such as the use of carcinogenic (cancer-causing) substances. A case in point is the raw material for PVC manufacture, chloroethene (vinylchloride monomer), which was widely used for several years before its hazardous nature became apparent. Ask your teacher for some reading references.

Examples of articles made from PVC.

Expanded polystyrene packaging.

As you may have gathered from your reading, there are often several ways of making a particular product. Some of the factors that a chemical company might take into account in deciding which route to use for a particular product are:

- availability of raw materials,
- energy needed for the process,
- hazards to the workers,
- environmental pollution.

The costs of raw materials, energy and labour vary from country to country and from time to time. Also variable are government regulations, trade union strength and social attitudes to hazards and pollution. Consequently, different processes may be in use for the manufacture of the same product according to the conditions prevailing when the plant was planned.

In ILPAC 8, Functional Groups, in the section on industrial processes we considered the factors involved in setting up chemical processes in general. You should refresh your memory on this before attempting to answer questions on the industrial preparation of polymers.

The formation of addition polymers usually requires the presence of a small amount of very reactive substance called an initiator. Peroxides are frequently used as initiators. They release radicals to initiate the free radical chain reaction. The mechanism for this was described in ILPAC 5, Introduction to Organic Chemistry, page 41, which you may wish to refer to if this is a requirement of your particular syllabus.

As well as free radical polymerisation there is also ionic polymerisation. The setting of superglue which you looked at in Exercise 72 is an example of this type.

Altering the conditions of polymerisation can lead to products that are physically quite different. A good example of this is in the industrial manufacture of high- and low-density poly(ethene) which you consider in the next exercise.

EXERCISE 91

Answers on page 197

Poly(ethene) is manufactured by two distinct processes and produces either high- or low-density poly(ethene).
a State the conditions for carrying out each process.
b Relate the difference in physical properties of high- and low-density poly(ethene) to their structure.
c Which of these processes is by a free radical mechanism?
d Which type of poly(ethene) would you use to make the following. Give reasons for your choice.
  i) Plastic carrier bags.
  ii) Medical equipment used in operating theatres.
  iii) Detergent (squeeze) bottle.
  iv) Bleach bottle.

The teacher-marked exercise that follows is about two alternative routes for PVC production. It is the sort of question where several different answers may be acceptable provided that the reasoning is sound. We suggest that you discuss the questions with other people in the group, and with your teacher. After you have sorted out your own ideas, write your answer to the question and hand it in to your teacher in the usual way.

EXERCISE

Teacher-marked

Read the following account and answer the questions.

Chloroethene (vinyl chloride), used for the production of poly(chloroethene) (polyvinyl chloride) plastics, can be manufactured from ethyne by reaction with hydrogen chloride in the presence of mercury(II) chloride as catalyst.

$$CH \equiv CH + HCl \xrightarrow[200\,°C]{HgCl_2} CH_2 = CHCl$$

The reaction is highly exothermic and temperature control is essential. Ethyne is obtained by the hydrolysis of calcium dicarbide; the latter is made by heating a mixture of coke and calcium oxide in an electric furnace at 2000 °C.

The ethyne process has now been largely superseded by a route based on ethene. Thus, ethene and chlorine react to form 1,2-dichloroethane which is also used as a solvent for the following reaction.

$$CH_2 = CH_2 + Cl_2 \xrightarrow{50°C} CH_2ClCH_2Cl$$

The product of this reaction is then decomposed by passing the vapour over a heated catalyst.

$$CH_2ClCH_2Cl \xrightarrow{500\ °C} CH_2 = CHCl + HCl$$

The hydrogen chloride is re-used in an oxychlorination process yielding 1,2-dichloroethane.

$$CH_2 = CH_2 + \tfrac{1}{2}O_2 + 2HCl \rightarrow CH_2ClCH_2Cl + H_2O$$

Ethene is readily available from petroleum by cracking and chlorine is obtained by the electrolysis of brine.

It has been found that prolonged exposure to chloroethene can lead to cancer of the liver.

a For each process, identify as many advantageous and disadvantageous features as you can.

b Give a reason why the second route has superseded the first.

c Under what circumstances could the ethyne route become relatively more competitive?

d List some of the precautions that have to be taken during the production and use of chloroethene.

e What are the raw materials needed for the ethene-based route and how will this effect the siting of the plant?

 f Draw a diagram to illustrate the ethene-based process for the production of poly(chloroethene). Use a generalised flow diagram from ILPAC 8, Functional Groups, page 40, as a guide.

g Draw a diagram to illustrate the structure of poly(chloroethene) including at least three monomer (repeat) units.

The next exercise concerns the production of poly(phenylethene) (polystyrene).

EXERCISE 92
Answers on page 197

The following reaction scheme shows a route for making phenylethene (styrene):

$$\bigcirc + CH_2 = CH_2 \xrightarrow{(1)} \bigcirc\!\!-CH_2CH_3 \xrightarrow{(2)} \bigcirc\!\!-CH = CH_2 + H_2$$

a State the types of reaction taking place at (1) and (2).

b Give the conditions needed (catalyst and temperature) for each stage of the reaction.

c Finally, the styrene is polymerised by a free radical process.
   i) Name a suitable initiator for this reaction.
   ii) Draw a diagram to illustrate the structure of a poly(phenylethene) molecule, including at least two monomer units.

To sum up your work on polymers and help you consolidate your knowledge you should now attempt one of the questions in the following teacher-marked exercise. They are all examination questions; ask your teacher to help you decide which is most appropriate for you.

EXERCISE
Teacher-marked

Choose **one** of the following questions, 1–3, and then attempt question 4.
1. Write an essay on 'synthetic polymers'. Aspects which you might consider include:
   a the nature of the chemical reactions used to make the polymers,
   b the methods used for their fabrication,
   c some specific applications of polymers, and
   d the effect of structure on the properties of a polymer.

2. **a** Give one example of, and explain the constitution of:
   i)   a polyalkene,
   ii) a polyester,
   iii) a polyamide.
   b Describe briefly how one of the above polymers is manufactured, mentioning the raw materials that are used at the start.
   c State, with your reasons, which one of the polymers in (a) you would choose as a suitable material for lining a container for aqueous alkalis, and why the other two are unsuitable. Give for each of the latter an important everyday application.
   d Illustrate, with the substances you have chosen in (a) the difference between addition polymerisation and condensation polymerisation.

3. Explain the chemical nature of the polymers polythene, Terylene and nylon-6 and of the processes involved in their preparation from the appropriate monomers. Is the distinction between addition and condensation polymerisation a useful one?

   Account for the general chemical inertness of polythene and indicate what reactions it might be expected to undergo. What bearing does this have on:
   a the usefulness of polythene and
   b the disposal of waste polythene?

   A polymer produced from a benzenedicarboxylic acid and ethane-1,2-diol can be used as a textile fibre; in what ways are the physical properties and uses of the polymer likely to be altered if the ethane-1,2-diol is replaced by propane-1,2,3-triol (glycerol)?

4.

Table 16

Polymer	Melting temperature/°C	Tensile strength /MPa	Elongation at fracture /%	Glass* temperature/°C	Relative cost
Low-density poly(ethene)	120	15	600	−20	1.5
High-density poly(ethene)	130	29	350	−20	1.5
Poly(phenyl-ethene)		40	2.5	100	1.5
Nylon-66	265	83	200	57	5
Phenolic resin (Bakelite)		50	0.6		1.0

*Glass temperature – this is the temperature at which the polymer chains in the amorphous region gain sufficient mobility to allow some softening of the plastic.

With reference to the data above, suggest reasons for the differences in
a i)   the melting points,
   ii) tensile strengths,
   of high- and low-density poly(ethene),
b the elongation at fracture of nylon-66 and phenolic resin,
c the relative cost of phenolic resin, low-density poly(ethene) and nylon-66.

The advantageous properties of plastics, in particular their inertness, causes problems of disposal once their useful life has finished. The most resource-efficient way to deal with waste plastics is to recycle them. This forms the subject of the next section. Check with your teacher whether you should cover this. If not, proceed to page 109.

■ 6.13 Plastics, the environment and recycling

OBJECTIVES

When you have finished this section you should be able to:
■ discuss the difficulties of the disposal of plastics through **non-biodegradability** and harmful combustion products;
■ describe the current methods for **recycling** plastics and the latest development of **biodegradable plastics**.

Find out from your local district council if they operate a plastics recycling centre in your area; they may produce a booklet on recycling facilities. Read about current methods of recycling plastics and the latest development of biodegradable plastics. Your teacher should be able to recommend some articles from magazines which will help you with the next teacher-marked exercise.

EXERCISE
Teacher-marked

a What properties of plastics makes them difficult to dispose of?
b Write a report or hold a class discussion that critically discusses the following statement in relation to recycling plastics.

'It is very important when considering recycling operations that the end-use really has a value. It is not very helpful to the environment or conservation of resource if the very process of recycling uses up more oil, energy and resource than it saves. The overall environmental equation has to balance for this to be worthwhile.'

Table 17
Typical recoverable energy values in MJ kg^{-1}

Material	Recoverable energy
Paper	14
Wood	16
PVC	22
PET (see Exercise 93)	22
Coal	29
Recycled mixed plastic bags	40
Polystyrene	40
Polyethylene	43.3
Polypropylene	43.3
Heating oils	44

Examples of articles made from recycled PVC.

A plastic bottle bank.

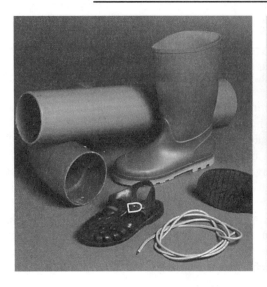

Mixed plastic waste.

Some of the following facts could add weight to your argument:
- Plastics take up 7–8% by weight, 12–15% by volume in excavated landfills.
- Most plastics have recoverable energy values similar to soft coal (see Table 17).
- Recycling of domestic used plastics has not yet been very successful because of the difficulties associated with collection and segregation. The cost involved in manual sorting outweighs the savings in material.
- A new company called REPRISE Plastics has installed a plant that will separate poly(chloroethene) (polyvinylchloride, PVC), poly(ethylene terephthalate) (PET) and poly(ethene) bottles collected from public sources. Poly(ethene) is separated by water flotation as PVC and PET are both more dense and will sink. The PVC and PET are then separated by an X-ray detection device which recognises the chlorine in PVC and activates a separation procedure.
- End-uses for recycled PVC are sewage pipes, cable coverings, shoe soles and anti-noise walls.
- A UK polystyrene recycling organisation targeting fast-food outlets uses recycled polystyrene for thermal insulation in homes.
- Mixtures of plastics can be used without separation. After melting it can be extruded into long timber-like sections for the building industry.

There have been recent moves to adopt some kind of coding system on packaging products which identify the material for ease of recycling. Currently some British companies are using their own coding systems which promotes inconsistency and confusion for the general public. The major trade associations are promoting an initiative to adopt the SPI code, originated by the Society of Plastics Industry Inc in the United States of America, which has already been widely adopted elsewhere in Europe.

In the next exercise we give you this coding system to identify different plastics and an opportunity to find out some of their uses.

EXERCISE 93

Answers on page 198

The SPI code for plastics is shown in Table 18. Complete a copy of this table. Identify as many of them as you can and as many of their uses as possible.

Table 18

SPI code	Systematic name	Common name	Uses
1 PETE			
2 HDPE			
3 V			
4 LDPE			
5 PP			
6 PS			

It would be quite interesting to see how many of these codes you could find on plastic packaging and bottles around your home.

Another option is to develop plastics that break down after disposal; these we call biodegradable plastics. This is the topic of the next exercise. The first part of the question requires you to write a summary. This type of question is becoming popular with examination boards. Check to see if you will be set this type of question in your final exam. Whether it is or not, it is a useful skill to practise. We gave a few tips on how to tackle this type of exercise in ILPAC 5, Introduction to Organic Chemistry, page 76. You may wish to refer to these before starting.

EXERCISE 94

Answers on page 198

Read the following passage about Biopol straight through, and then more carefully, in order to answer the following questions.

a Write a summary in continuous prose, in no more than 150 words, to describe the current manufacture, properties and uses of Biopol.
Credit will be given for answers written in good English, using complete sentences and with the correct use of technical words. Numbers count as one word, as do standard abbreviations and hyphenated words. If you include chemical equations or formulae they do not count in the word total, nor does the title to your account. At the end of your account, state clearly the number of words you have used. There are penalties for the use of words in excess of 150.

b i) Draw structural formulae for 3-hydroxybutanoic acid and 3-hydroxypentanoic acid. Show how one molecule of each could join together to form part of the polyester Biopol.
ii) Predict whether the polymer Biopol would be a thermosetting or a thermoplastic polymer. Explain your answer.

Items made from Biopol.

Biopol: a New Biodegradable Plastic

Since the early 1950s, polymeric materials have revolutionised the way most people live in industrialised countries. Because of their versatility and relatively low cost, polymers are essential components of most consumer articles, with the total world production now in excess of 100 million tonnes per annum.

The development of new products has always been high on the list of priorities for the chemical industry. There are two main reasons for this: firstly to increase the UK share of the highly competitive world plastics market; and secondly to combat the environmental problems associated with the growth of this market. Tough, durable, throw-away plastic products are difficult to dispose of – do you, for instance, bury them in a landfill site, or do you burn them in an incinerator? With the level of environmental concern at an all-time high, the general public is calling for materials with the toughness and flexibility of plastics, yet with the environmental 'friendliness' of non-plastics – that is, materials that are biodegradable.

Recently, ICI announced the launch of a new biodegradable plastic – 'Biopol'. This new plastic has all the durability, stability and water resistance of conventional plastics, but on disposal it is quickly and efficiently broken down into carbon dioxide and water.

'Biopol' polymers are linear polyesters of 3-hydroxybutanoic acid and 3-hydroxy-pentanoic acid. These acids are produced by the fermentation of a mixed feedstock of carbohydrate and organic acid by the naturally occurring bacterium *Alcaligenes eutrophus* which is widely distributed in soil, and fresh and salt waters, and is involved directly in the spoilage of food.

In the first stage of production, *Alcaligenes eutrophus* is inoculated into a growth medium containing glucose and mineral salts, at the specified temperature for optimum growth. During a period of rapid growth, biosynthesis of the polymer begins. Although it is strictly not a lipid, the polymer will be a component of the so-called lipid granules of the microorganism. At the end of the growth cycle the organisms have accumulated up to 80% of their dry mass as the polymer. The cells are then harvested and treated so that they break open. The crude polymer is removed by solvent extraction and then purified. What is in fact an energy reserve for the microorganism, like fat in mammals, now becomes a commercial plastic.

How does the microorganism produce the necessary monomers to form the polymer? As well as 2-hydroxypropanoic acid (lactic acid) being formed as a product of the fermentation of carbohydrates, other products are possible including 3-hydroxybutanoic acid. Chemists, by adding pentanoic acid to the growth medium, have produced Biopol with a range of compositions from 0–20% of 3-hydroxypentanoic acid.

Biopol has a variety of uses depending on its composition. Since it is a biocompatible compound, it can be used in a range of medical implants, e.g. by choosing the correct composition of the polymer, it is possible to produce slow release capsules for veterinary use, with the polymeric outer material slowly breaking down to release controlled amounts of drugs at the site of implant.

Biopol products can be made using conventional polymer-processing technology, although care must be taken at all times because, as a polyester, Biopol is liable to be unstable at high temperatures. At the typical processing temperatures used for Biopol (185–190 °C) the time the polymer is in the melt needs to be minimised. At temperatures above 205 °C the polymer degrades rapidly.

Although we will probably be using Biopol in a variety of ways in the future, most available literature concentrates on its biodegradability. So why does Biopol biodegrade so easily? For an answer to this we must remember that the substance is a polyester and contains compounds that are present in a range of living systems. On disposal to a suitable site and with the correct conditions, fungi and bacteria break down the Biopol in a few weeks. The first step is probably the hydrolysis of the ester linkages. After hydrolysis, the compounds produced are metabolised through a fatty acid oxidation cycle producing energy, carbon dioxide and water.

Will Biopol ever be cheap enough to replace non-degradable plastics such as poly(ethene)? The answer to this must be no. However, with effective 'green' advertising, consumers are probably prepared to pay a premium for this material. Also the gene responsible for synthesising poly-3-hydroxybutanoic acid was identified in 1987 and since then scientists have been attempting to introduce it into other organisms. It is possible that crop plants with this gene may soon be able to produce the polymer on a large scale, so reducing its cost.

Is the widespread adoption of biodegradable plastics the long-term answer for a waste-making industrialised society? What about the recycling of plastics? The answers to these questions lie in the future, but the production of Biopol is a step in the right direction.

(801 words)

(Adapted from an article by Dr. P. S. Phillips, *Education in Chemistry*, Volume 28, No. 5 September, 1991).

DO NOT FORGET TO ANSWER BOTH PART (*a*) AND PART (*b*).

Now we look at some of the methods used to determine the structure of organic compounds.

7 DETERMINATION OF STRUCTURE

There are many tools available to a chemist for determining the structure of an organic molecule. Figure 42 outlines the stages and various methods that are available. The method adopted depends largely on the complexity of the compound concerned.

Figure 42

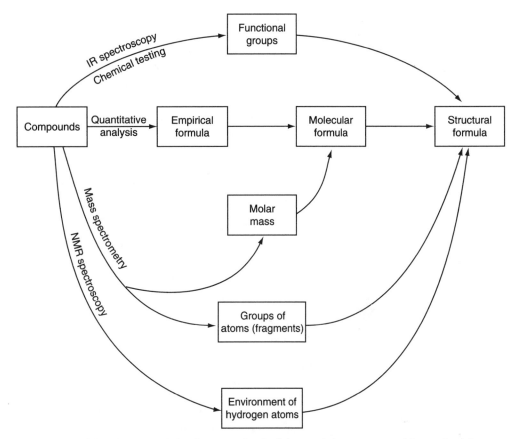

We first consider the more laborious method of determining structural formula using results from quantitative analysis, molar mass determination and chemical testing. Later, we show you the rather more sophisticated methods available using mass spectrometry, IR and NMR spectroscopy.

■ 7.1 Empirical formula from quantitative analysis

OBJECTIVE When you have finished this section you should be able to:
■ deduce the **empirical formula** of a compound from elemental percentage composition.

When an organic compound is burned in an excess of oxygen, the carbon is converted to carbon dioxide and the hydrogen to water. The amounts of carbon and hydrogen present used to be determined by passing the products of combustion into pre-weighed tubes containing substances that selectively absorb them. Now the combustion products are determined by measuring the thermal conductivity of the gas stream before and after absorption. Other methods must be used to determine nitrogen, sulphur and halogen content. The entire process is called 'quantitative analysis'.

Unless your teacher or syllabus states otherwise you will not be required to determine empirical formulae directly from combustion data but more simply from mass percentage data. We have included calculations using combustion data in Appendix 3 if it is a requirement in your particular syllabus.

■ 7.2 Calculating empirical formula from percentage composition by mass

We gave you a worked example on this type of problem in ILPAC 1, The Mole, page 51. You may wish to refresh your memory on this method before attempting the next exercise.

EXERCISE 95
Answer on page 200

Compound A was analysed and found to contain 52.2% carbon, 13.0% hydrogen and 34.8% oxygen. Determine its empirical formula.

Having established the empirical formula of a compound, the next step is to determine its molecular formula.

■ 7.3 Molecular formula from empirical formula

OBJECTIVE

When you have finished this section you should be able to:
■ calculate the **molecular formula** of a compound from empirical formula and molar mass.

In order to determine the molecular formula of a compound not only do we need to know its empirical formula, we also need to know its molar mass. This can be determined by a variety of methods. The preferred technique is mass spectrometry as it can give other information apart from molar mass. You may wish to refresh your memory on this technique from ILPAC 1, The Mole, page 51. We give a whole section on interpreting mass spectra in more detail later.

To give you an idea of how we use empirical formula and molar mass to determine molecular formula we give you a simple worked example.

WORKED EXAMPLE

The empirical formula of compound X is found from quantitative analysis to be C_2H_5. Mass spectrometry gave a relative molecular mass of 58. What is the molecular formula of X?

Solution

1. Determine the relative molecular mass corresponding to the empirical formula, C_2H_5.

$$M_r(C_2H_5) = (2 \times 12) + (5 \times 1) = 29$$

2. Substitute into the expression:
   molecular formula = (empirical formula)$_n$

$$\text{where } n = \frac{M_r(\text{compound X})}{M_r(C_2H_5)} = \frac{58}{29} = 2$$

$$\text{molecular formula} = (C_2H_5)_2 = \mathbf{C_4H_{10}}$$

You should now be able to do the next exercise.

EXERCISE 96
Answers on page 200

An organic compound A, has a relative molecular mass of 178 and contains 74.2% C, 7.9% H and 17.9% O. Determine:
a the empirical formula of A,
b the molecular formula of A.

Having established the molecular formula of a compound, we can now go on to determine its structural formula.

■ 7.4 Structural formula from molecular formula

OBJECTIVE   When you have finished this section you should be able to:
■ use the results of **chemical tests** together with molecular formula to deduce the
structure of a compound.

Before we can write the structural formula of a compound, we need additional
information about its chemical reactions. The following exercises, which are A-level
questions, include such information. In ILPAC 8, Functional Groups, Tables 22 and 23,
pages 111–114 we give characteristic tests for these compounds. Your knowledge of the
chemistry of functional groups and reference to these tables should enable you to write a
structural formula for each of the compounds under investigation.

EXERCISE 97
Answers on page 201

Two substances, Q and R, each have the elemental composition C = 60.0%, H = 13.3%,
O = 26.7% (by mass) and a relative molecular mass of 60.1.
a Calculate the empirical formula of Q and R.
b What is their molecular formula?
With sodium, Q was unreactive but R produced a flammable gas. R also reacted with
warm, acidified potassium dichromate, giving a product which, when isolated by
distillation, gave a precipitate of silver when warmed with aqueous silver nitrate, sodium
hydroxide and ammonia (Tollens reagent).
c Identify Q and R.
d Write simplified equations for the reactions of R as described above.

EXERCISE 98
Answers on page 202

A liquid X contains by mass 31.5% of carbon, 5.3% of hydrogen and 63.2% of oxygen.
Its solution in water liberates carbon dioxide from sodium carbonate and from the
resulting solution a substance of formula $C_2H_3O_3Na$ can be obtained. With phosphorus
pentachloride, X gives hydrogen chloride and a compound of molecular formula
$C_2H_2OCl_2$.
State and explain the information that may be obtained from each of these facts. Give
the molecular formula of X and devise a synthesis of it from ethene (ethylene).

EXERCISE 99
Answers on page 203

a Compound A is a liquid hydrocarbon of molecular weight (relative molecular mass)
78 and contains 92.3% carbon. On treatment with concentrated nitric acid in the
presence of concentrated sulphuric acid, A forms compound B which, upon reduction,
yields compound C containing carbon, hydrogen and nitrogen only. Deduce the
identity of compounds A, B and C, and explain the function of the concentrated
sulphuric acid in the conversion of A into B.
b In a subsequent experiment, a sample of C was dissolved in hydrochloric acid and,
after cooling in iced water, treated with aqueous sodium nitrite solution. Half of this
mixture was heated to above 60 °C whereby a gas was seen to be evolved. After
cooling, this solution was added to the other half of the original mixture whereupon a
coloured precipitate was formed. Account as fully as you can for the observations
described above. (H = 1.0, C = 12.0)

Sometimes you may have to calculate an empirical formula and use it to help you work
out a more difficult synthetic pathway. After doing the calculation in the next exercise,
you may find it useful to summarise the information in a series of steps and proceed as in
the worked example in ILPAC 8, on page 204.

EXERCISE 100
Answers on page 204

A compound, P, contains carbon 59.4%, hydrogen 10.9%, nitrogen 13.9% and oxygen 15.8%, by mass. (Relative atomic masses: H = 1, C = 12, N = 14, O = 16.)

On distilling P with hot dilute aqueous sodium hydroxide, Q, C_2H_7N, is formed immediately and R, $C_3H_6O_2$, when the residual solution is acidified. A colourless gas and a liquid, S, are formed on treating Q with an ice-cold solution of sodium nitrite and dilute sulphuric acid. On warming with S and a few drops of concentrated sulphuric acid, R forms a sweet-smelling liquid, T.

Identify P, Q, R, S and T and write equations for the reactions described.

EXERCISE 101
Answer on page 205

A chemist, asked to identify an organic compound labelled X, carried out a series of experiments and reported that:

The composition of X by mass is C 66.7%, H 11.1%, O 22.2%.
The molar mass of X is 72.
X reacts with a solution of the reagent 2,4-dinitrophenylhydrazine to give an orange precipitate.

Use the information in the report to find:
a the empirical formula of X,
b the molecular formula of X,
c the name of the functional group in X,
d two possible structural formulae for X.

The chemical methods for determining structural formulae can be very laborious and time consuming, especially for complex compounds. Physical methods are now more often employed: these can be done very rapidly but they require the use of expensive apparatus rarely found in schools.

■ 7.5 Physical methods for determining structure

We shall cover the techniques of mass spectrometry, infra-red and nuclear magnetic resonance spectroscopy. Before spectroscopy was commercially available, identification of an unknown substance required breaking it down chemically into its constituent elements to find its empirical formula, and using specific reactions to determine its structure; a time-consuming and often inaccurate procedure. Spectroscopy has shortened the time required to determine the structure of a compound considerably.

All the techniques contribute valuable information about the structure of organic compounds. Structure determination therefore often involves the use of more than one of these methods. You will have the opportunity of using evidence from up to three different types of spectra to suggest probable structures for given compounds later. Since syllabus content varies quite considerably on this particular topic you should check with your teacher which sections you should cover. First we look at infra-red spectroscopy.

Sections on these techniques are included in the ILPAC video programme 'Instrumental techniques' and the Royal Society of Chemistry video 'Modern chemical techniques'. If the programmes are available, view these sections now or at some other convenient time.

■ 7.6 Infra-red (IR) spectroscopy

This is a very useful technique for determining the presence of functional groups in a molecule or for confirming the identity of a compound by comparing its infra-red spectrum with that of a known compound.

Infra-red spectra arise principally as a result of absorption of energy at particular wavelengths corresponding to the changes in vibrations of bonds within the molecules under investigation. These vibrations can be due either to stretching or bending of bonds, as shown in Fig. 43.

Figure 43
Different modes of vibration.

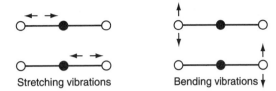

Stretching vibrations   Bending vibrations ↓

Read about the general features of the infra-red spectrometer (source, cell and method of recording). Find out how this technique can be used to identify the functional groups present in a compound.

Note that some books identify the absorbed radiation by its wavelength, and others by reciprocal wavelength, called wavenumber, the most common units being m (metres) and cm^{-1} respectively. Listed values in these units are easily interchangeable; one is simply the reciprocal of the other with the appropriate conversion factor. For example, one book lists the wavelength of maximum absorption by a C—H bond as 3.4×10^{-6} m, while another gives the wavenumber, 2940 cm^{-1}:

$$\text{i.e.} \quad \frac{1}{3.4 \times 10^{-6}\,\text{m}} = 2.94 \times 10^{5}\,\text{m}^{-1} = 2940\,\text{cm}^{-1}$$

In the exercise that follows, you can see how inspection of a series of infra-red spectra of related compounds enables you to associate the absorption of energy at particular wavelengths with the presence of particular groups of atoms. High values of transmission (or transmittance) of say over 90% mean that the sample is absorbing little or no energy. Dips or troughs in the graph (confusingly often referred to as 'peaks') means that energy is being absorbed.

EXERCISE 102
Answer on page 205

Figure 44
Infra-red absorption spectra of decane, trichloromethane and tetrachloromethane.

Figure 44 shows the infra-red absorption spectra of decane, trichloromethane and tetrachloromethane. Study them and attempt the question which follows.

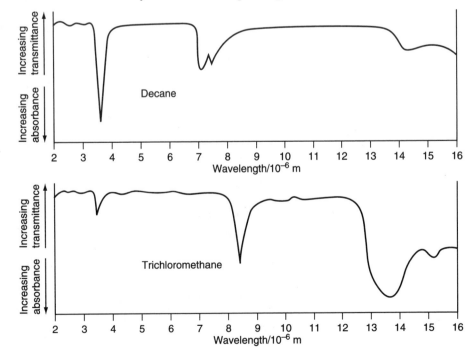

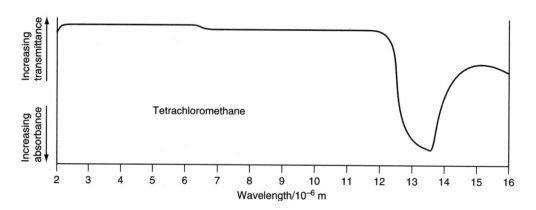

Deduce the wavelengths at which the C—C, C—H and C—Cl bonds absorb radiation.

By comparing the infra-red spectra of related compounds, it is possible to construct lists of characteristic absorption wavelengths. One such list is shown in Fig. 45.

The characteristic absorption wavelength given by a particular bond not only depends on the two atoms bonded, but also on the effect of neighbouring atoms. For example, the C—H bond absorbs at slightly different wavelengths in different compounds. This means that it is impossible to assign a precise absorption wavelength to a bond. Nevertheless, as you can see in Fig. 45, we can identify **bands** corresponding to bonds in functional groups.

The spectrum of a particular compound can be quite complex; we shall concentrate our attention on the main absorption troughs (or peaks, as they are sometimes called).

Figure 45
Characteristic infra-red absorptions for functional groups.

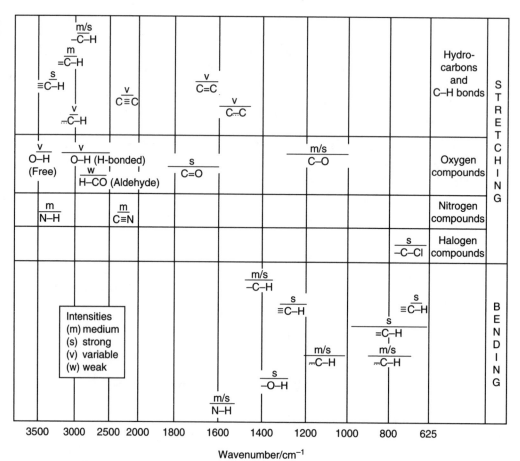

You can now use Fig. 45 to identify the bonds that are responsible for some of the absorption bands shown in some infra-red spectra.

EXERCISE 103

Answer on page 206

Below is the infra-red spectrum of phenylamine, $C_6H_5NH_2$. Study it and, after consulting Fig. 45, identify the bonds that are likely to be responsible for the absorptions at **A**, **B**, **C** and **D**.

Figure 46
Infra-red absorption spectrum of phenylamine.

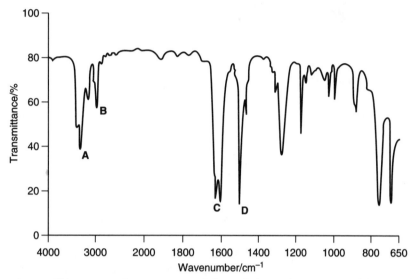

It is rarely possible to correlate every peak in an infra-red spectrum with a particular molecular vibration, but usually, specific bonds can be identified as you can see in the next exercise.

EXERCISE 104

Answer on page 206

Below are the infra-red spectra of benzene and ethyl ethanoate. Study them and, after consulting Fig. 45, decide which is which. Explain your reasoning.

Figure 47
Infra-red spectra.

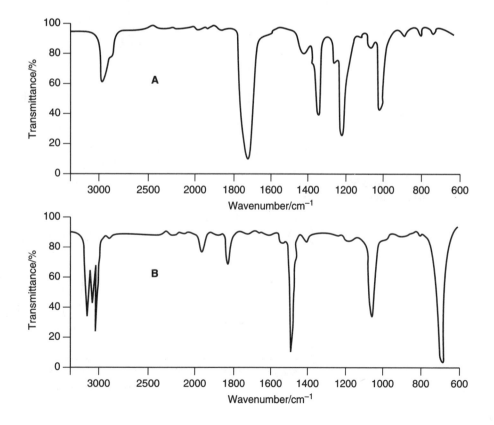

You should now attempt the following teacher-marked exercise which requires you to identify three compounds from IR spectra as a group activity with some discussion.

EXERCISE

Teacher-marked

The IR spectra of three compounds butanal (C_3H_7CHO), propanoic acid ($CH_3CH_2CO_2H$) and butan-1-ol (C_4H_9OH) are shown below. Using data from Fig. 45, identify which spectrum corresponds to which compound. Give full reasons for your decisions. (Clue: try not to confuse the narrow dip at 3000 cm^3 for C—H stretching with the much broader dip caused by an O—H group.)

Figure 48
IR spectrum for compound A.

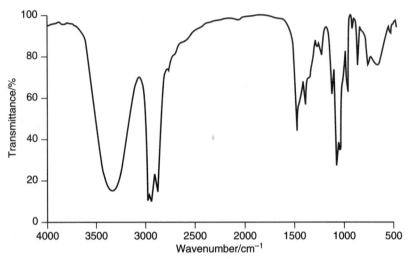

Figure 49
IR spectrum for compound B.

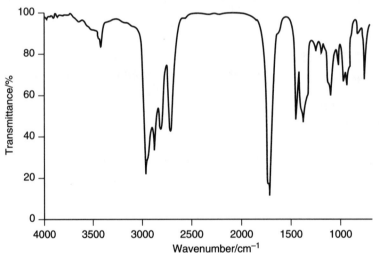

Figure 50
IR spectrum for compound C.

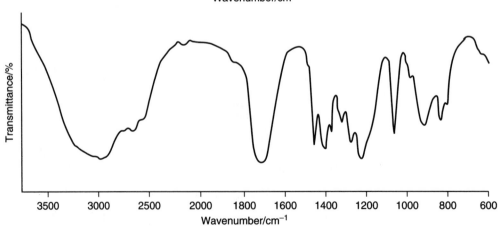

You should now attempt the following two exercises which are examination questions.

EXERCISE 105

Answers on page 206

Figure 51

The infra-red spectrum shown in Fig. 51 was obtained from a compound X of formula $C_5H_7O_2N$.

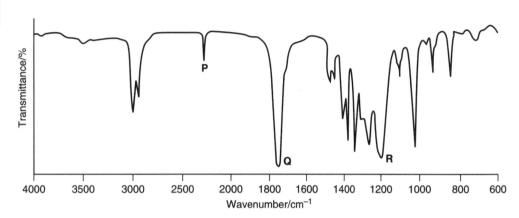

Identify the bonds responsible for the absorptions, **P**, **Q** and **R**, and hence suggest a structure for X.

(Hint: the trough for an —OH group is much broader than for a C—H group. See Fig. 45.)

EXERCISE 106

Answer on page 206

Figure 52

Ethanol (CH_3CH_2OH), ethanoic acid (CH_3CO_2H) and propanone (CH_3COCH_3) may be distinguished by their infra-red spectra. Figure 52 represents the IR spectrum of one of these. Decide which compound gave the spectrum and explain your choice.

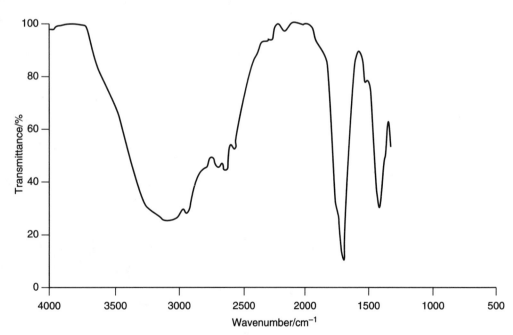

Determine which one it is by using the data in Fig. 45.

We now consider some other applications of infra-red spectroscopy.

■ 7.7 Applications of IR spectroscopy

One application of infra-red absorption spectroscopy is in the analysis of breath samples given by drivers using an instrument called an **intoximeter**. It is used in police stations to back up roadside breath tests. The results can be used as evidence in court. Figure 53 illustrates how this instrument works and Fig. 54 shows an IR spectrum for ethanol. Study them and attempt the exercise that follows.

Figure 53

The **intoximeter** is used in police stations to back up the road side breath tests. It works by measuring the infra-red light absorbed by ethanol. The results can be used as evidence in court

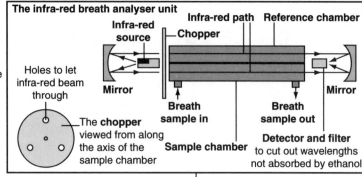

The **infra-red breath analyser unit**

Infra-red source

Infra-red path

Reference chamber

Chopper

Mirror

Holes to let infra-red beam through

The **chopper** viewed from along the axis of the sample chamber

Breath sample in

Sample chamber

Breath sample out

Mirror

Detector and filter to cut out wavelengths not absorbed by ethanol

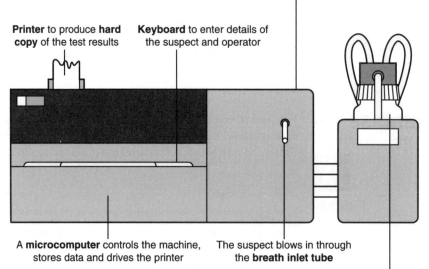

Printer to produce **hard copy** of the test results

Keyboard to enter details of the suspect and operator

A **microcomputer** controls the machine, stores data and drives the printer

The suspect blows in through the **breath inlet tube**

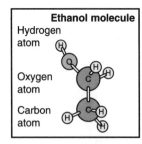

Ethanol molecule

Hydrogen atom

Oxygen atom

Carbon atom

In an **ethanol molecule** it is the vibrations of the carbon–hydrogen bonds which make it absorb infra-red radiation of wavelength $3.39\,\mu m$

A **propanone molecule** can be found in the breath of diabetics. It also has carbon–hydrogen bonds and could interfere with a breath test for ethanol

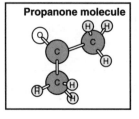

Propanone molecule

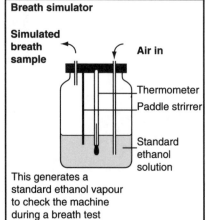

Breath simulator

Simulated breath sample

Air in

Thermometer

Paddle stirrer

Standard ethanol solution

This generates a standard ethanol vapour to check the machine during a breath test

Figure 54
IR spectrum for ethanol.

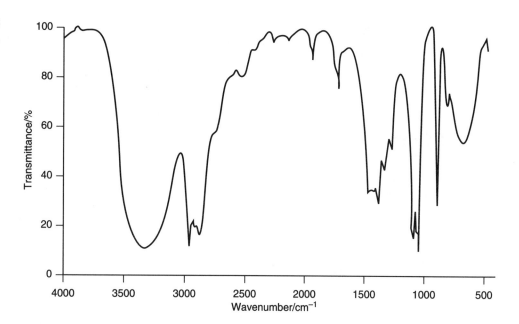

EXERCISE 107

Answers on page 206

a Determine the bond vibrations responsible for the major absorption at 3340 cm^{-1} and 2950 cm^{-1}.

b Only one of the above absorptions can be used to determine the concentration of ethanol. Decide which you think it is, with reasons.

c People with diabetes sometimes produce propanone in their breath. Why is it important that the intoximeter is designed to eliminate the contribution to the IR absorption by propanone (acetone).

Another application for this technique is in monitoring air pollution at various sites around the country. Some of the data was used in various exercises in the topic of air pollution in ILPAC 5, Introduction to Organic Chemistry.

We now study another physical method of determining structural formulae – mass spectrometry.

■ 7.8 Mass spectrometry in organic chemistry

You have already studied the mass spectrometer in ILPAC 1, Atomic Structure, where you considered its use in determining relative atomic masses. In the same way, the mass spectrometer can be used to determine the relative molecular mass of a compound but, in addition, the structure can be determined by identifying the fragments formed when the molecule is ionised.

OBJECTIVES

When you have finished this section you should be able to:

■ identify the **parent molecular ion** and **isotopic molecular ion** from a **mass spectrum**:

■ explain the significance of some of the major peaks in a mass spectrum;

■ explain why each major peak in a mass spectrum is accompanied by several **smaller peaks**;

■ explain the use of the (M + 1) peak in a mass spectrum for determining the number of carbon atoms in organic molecules;

■ explain the use of mass spectrometry in **isotopic labelling** to determine a reaction pathway.

First, refresh your memory on how a mass spectrometer works and how to identify peaks on a simple mass spectrum by reading Sections 9.1 and 9.2 of ILPAC 1, Atomic Structure. Read the section of your textbook that deals with the application of the mass spectrometer in determining relative molecular masses using the mass/charge ratio (m/e) and elucidating the structure of organic compounds. You should then be able to do the following exercises.

EXERCISE 108

Answers on page 206

Figure 55
Mass spectrum of ethanol.

Figure 55 shows the mass spectrum of ethanol. Study it and try the questions that follow.

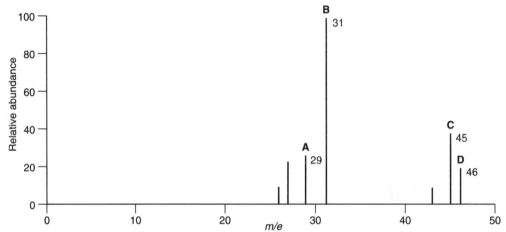

a How can a single substance such as ethanol give several peaks on the mass spectrum?
b Write the molecular formulae for the species giving traces at **A**, **B**, **C** and **D**.

In the last exercise, you saw how the highest mass/charge ratio in the spectrum corresponds to the single-positively charged ion of the complete molecule, known as the molecular ion, M^+. However, in some spectra there is a very small peak at $(M + 1)$ due to the presence of molecules containing the ^{13}C isotope, as shown in Fig. 56.

Figure 56

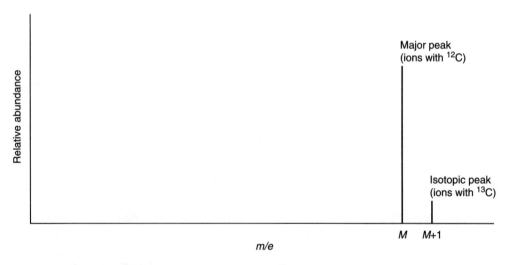

The isotopic peak is relatively small because the ^{13}C isotope has a natural abundance of only 1%, i.e. one ^{13}C atom for every 99 ^{12}C atoms. The more carbon atoms there are in a molecule the greater the chance that *one* of them will be a ^{13}C atom. The chances that two (or more) are ^{13}C also increases of course but is nearly always negligibly small.

Generally, for an organic compound containing N carbon atoms, the height of the $(M + 1)$ peak is approximately $N\%$ of the height of the M peak. With this in mind, attempt the next exercise.

EXERCISE 109

Answers on page 207

a Carbon consists of 99% of the isotope ^{12}C and 1% of the isotope ^{13}C. One way of estimating the number of carbon atoms in the molecule of a compound is to examine the mass spectrometer peaks which the molecule gives. The peak corresponding to the second highest mass is caused by the molecule ion having only ^{12}C in it but the highest mass peak is due to a molecule ion with one ^{13}C atom in it.

A hydrocarbon X gives a peak at mass M and a smaller peak at mass $(M + 1)$. There are no significant peaks at higher mass than this. The peak at mass M is 12.5 times as intense as the peak at mass $(M + 1)$.

X contains 92.4% carbon by mass. X decolorises bromine water and one mole of X reacts with one mole of bromine molecules. What is the molecular formula and probable structure of X?

b Bromine consists of a mixture of isotopes ^{79}Br and ^{81}Br. Assuming no bonds are broken in the mass spectrometer and that hydrogen has only one significant isotope, how many peaks will be given by the compound CH_2Br_2 and what will each peak be due to?

In the next exercise, you look at the spectra of two hydrocarbons and explain why each peak is accompanied by several smaller peaks.

EXERCISE 110

Answers on page 207

Figure 57
Mass spectra of two hydrocarbons.

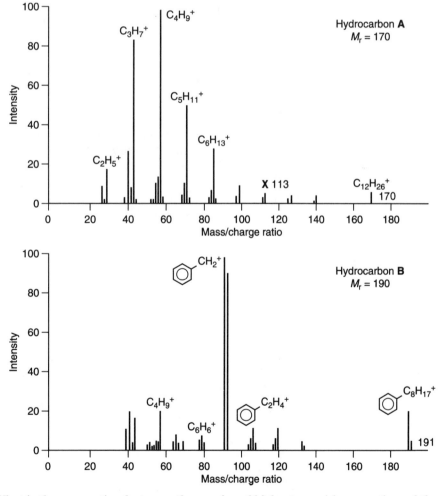

a What is the connection between the species of highest mass/charge ratio and the substance being analysed?

b Explain why each peak is accompanied by several smaller peaks.

c What is the probable formula for the species giving the trace marked **X** in the spectrogram for **A**?

EXERCISE 111
Answer on page 208

Identify the minor peaks associated with the major peak $C_5H_{11}^+$, shown in Fig. 57.

EXERCISE 112
Answers on page 208

Figure 58

Examine the mass spectrum of chlorobenzene shown in Fig. 58 and answer the questions that follow.

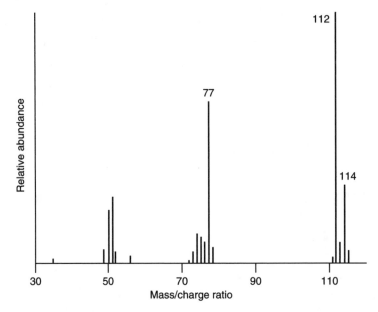

Mass/charge ratio

a Identify and explain the presence of two molecular ion peaks.
b Suggest an explanation for the presence of the ions of mass/charge ratios 113 and 115.
c Identify, with a reason, the fragment responsible for the peak of mass/charge ratio 77.

The mass/charge ratio for a particular ion can be measured very accurately indeed (high-resolution mass spectrometry) and thus can be used to obtain the molecular formula directly even when the component elements have not been fully identified. Consider, for instance, the peak at mass/charge ratio 170 due to the molecular ion $C_{12}H_{26}^+$ in Fig. 57. At high resolution, this could be distinguished from $C_{11}H_{22}O^+$ and $C_{11}H_{24}N^+$ because of small differences in mass between CH_4 and O, and between CH_2 and N.

$$M_r(C_{12}H_{26}) = 170.2035$$
$$M_r(C_{11}H_{22}O) = 170.1671$$
$$M_r(C_{11}H_{24}N) = 170.1909$$

Similarly, if a high-resolution spectrum gives the relative molecular mass as 170.1307, tables of all possible formulae show that the molecular formula is $C_{10}H_{18}O_2$. Fragments of closely similar mass can be distinguished in the same way, so you can see how powerful a technique mass spectrometry is.

The identification of fragment ions gives important clues to the structure of the parent molecule although it is not always easy to deduce a structure completely from a single spectrum. However, by studying the spectra of known compounds, the trained spectroscopist can easily recognise structural features and build up a model of the whole structure.

In the next teacher-marked exercise you use mass spectrometry in isotopic labelling to determine a reaction pathway. You have already encountered isotopic labelling in the esterification reaction between ethanol and ethanoic acid in ILPAC 8, Functional Groups, Exercise 10, page 17. You may wish to refresh your memory on this first.

EXERCISE
Teacher-marked

Methanol, CH_3OH, can be made to react with benzoic acid (benzenecarboxylic acid), $C_6H_5CO_2H$, to produce the ester methyl benzoate. In theory, there are two possible pathways for the reaction:

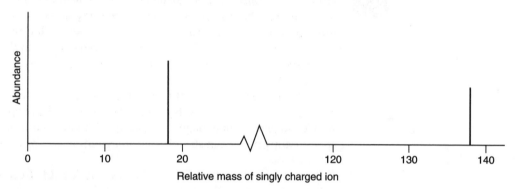

In Pathway 1, the oxygen atoms in the ester both come from the carboxylic acid.
In Pathway 2, only **one** of the oxygen atoms in the ester comes from the carboxylic acid.

In an experiment, in 1938, designed to discover which pathway is followed in practice, I Roberts and HC Urey at Columbia University, USA, reacted O–18 enriched methanol with benzoic acid. The ester and water produced were analysed by mass spectrometry. The mass spectrum below shows the relevant results.

Figure 59

```
Abundance

      |
      |              |
      |              |
      |              |                                              |
      |              |                                              |
      |              |                                              |
      |              |                                              |
      |              |                                              |
      |              |                 /\                           |
      |_____|_____/\/_____|____
      0      10      20          120        130        140
              Relative mass of singly charged ion
```

Identify the species giving rise to the peaks shown in the mass spectrum and hence, deduce which of the two possible pathways of the esterification is taken.

You now consider some of the other applications of mass spectrometry.

■ 7.9 Applications of mass spectrometry

There are other applications apart from the identification of organic compounds. Some of them are listed below.

Mass spectrometry has been used to:

■ determine drug abuse by athletes,
■ determine the presence of hallucinogens in blood,
■ identify bile acids in the gall bladder of an Egyptian mummy,
■ characterise tar from the Elizabethan warship *Mary Rose*,
■ detect dioxins and polychlorinated biphenyls (PCBs) as evidence of pollution,
■ identify the type of oil from an oil spill and trace it back to the tanker that dumped the oil.

Find out from suitable references suggested by your teacher about some of the applications of mass spectrometry listed above. Very few chemistry textbooks cover this topic. *Education Guardian*, 28 June 1994, pages 14–15, 'Detectives in the lab' is a very good article on this. Your school librarian or teacher may have back copies of this very useful resource.

We now study nuclear magnetic resonance spectroscopy.

■ 7.10 Nuclear magnetic resonance (NMR) spectroscopy

OBJECTIVES   When you have finished this section you should be able to:
■ explain how the chemical environment of a proton affects the magnetic field it experiences and hence the absorption of energy at resonance;
■ explain what is meant by **chemical shift**;
■ explain the use of the **δ scale** and **TMS** (tetramethylsilane) as a standard;
■ outline the applications of **NMR spectroscopy**.

 Read from resources suggested by your teacher a description of how NMR spectra give information concerning the structure of hydrogen-containing groups within a molecule.* Also find out about its use in medicine in body scanners. Since textbooks rarely cover this topic we lead you through the next exercises with rather more guidance and information than usual. Do not worry too much if you find the method of producing NMR spectra difficult to understand. Most examination questions focus on their interpretation.

NMR filled a 'gap' evident in X-ray diffraction results which did not show up hydrogen atoms at all well. Hydrogen nuclei have the property of spin (like electrons) which will give them a magnetic field. It will help in your understanding of NMR spectroscopy if you can stretch your imagination and picture these nuclei behaving like bar magnets.

If small bar magnets are placed in a magnetic field, the situation shown in **A** below will occur. Situation **B** could only happen if the smaller magnets are forced into that position against the repulsive force. They could stay in this position (like a ball on the top of a hill) but the slightest disturbance causes them to fall back to the lower energy state.

Figure 60

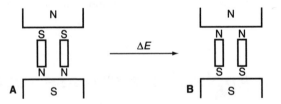

Magnetic nuclei aligned against the magnetic field have a higher energy

If we can imagine nuclei with spin behaving in a similar way, they will line up in a magnetic field as in **A**, i.e. the lower energy position. If sufficient energy is applied, however, a few nuclei will 'flip' into the higher energy position as in **B** above, absorbing energy ΔE which corresponds to a particular radio frequency (RF).

Figure 61 shows a simplified diagram of an NMR spectrometer. When the molecules under investigation absorb this energy in order to 'flip' the reorientation of nuclear spins, they induce a signal in a detector which is recorded as a resonance signal.

* Although several other nuclei can be determined by this technique, here we deal only with hydrogen nuclei (proton), and the term 'proton magnetic resonance' (pmr) spectroscopy is often used.

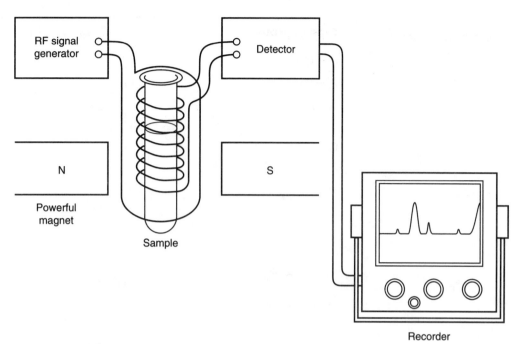

Figure 61
Simplified diagram of an
NMR spectrometer.

In practice, the radio frequency is usually kept constant and the magnetic field changed
slowly (i.e. scanned) using an electromagnet to see which values cause protons to 'flip'.
This is the basis of NMR spectroscopy. Since hydrogen nuclei (protons) in different
chemical environments absorb at slightly different applied fields the NMR spectrum will
show them up. You seek explanations for this in the next exercise.

EXERCISE 113

Answers on page 208

a What might protect (shield) a hydrogen nucleus (proton) from the external magnetic
field?

b How might this protection differ between hydrogen atoms bonded to electronegative
atoms such as oxygen and hydrogen atoms bonded to, say, carbon?

To summarise very briefly, hydrogen atoms in different chemical environments will
absorb radio waves at different frequencies. The amount of shielding is measured in
NMR spectra by chemical shift (δ) from a reference compound called tetramethylsilane
(TMS) which is arbitrarily given a chemical shift value of zero. The greater the chemical
shift, the less the shielding.

The reference
compound
for NMR

$$\begin{array}{c} CH_3 \\ | \\ H_3C-Si-CH_3 \\ | \\ CH_3 \end{array}$$

TMS

Table 19 gives typical chemical shift values (δ) relative to TMS = 0 for different types of
proton. (Since TMS = 0 is arbitrary we can have negative shift, e.g. in aromatics.) You
use it to identify different types of proton in the exercise that follows.

Table 19
Typical proton chemical shift
values (δ) relative to
T.M.S. = 0

Type of proton	Chemical shift (ppm)
R—CH$_3$	0.9
R—CH$_2$—R	1.3
R$_3$CH	2.0
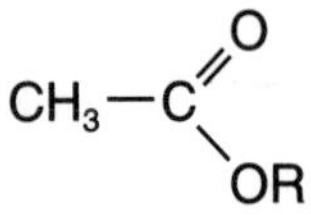 CH$_3$—C(=O)OR	2.0
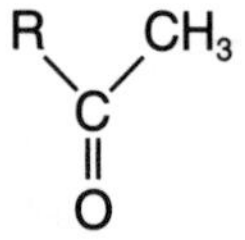 R\C(CH$_3$)=O (i.e. R–C(CH$_3$)=O)	2.1
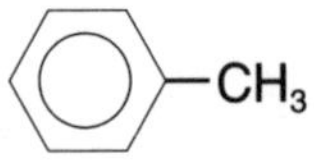 ⬡—CH$_3$	2.3
R—C≡C—H	2.6
R—CH$_2$—Hal	3.2–3.7
R—O—CH$_3$	3.8
R—O—H	4.5*
RHC=CH$_2$	4.9
RHC=CH$_2$	5.9
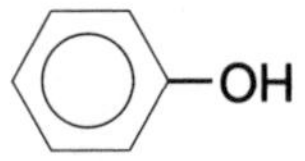 ⬡—OH	7*
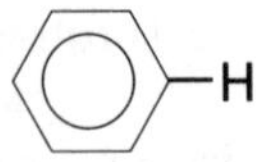 ⬡—H	7.3
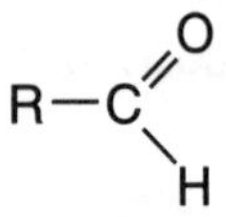 R—C(=O)H	9.7*
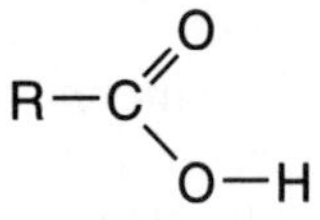 R—C(=O)O—H	11.5*

* Sensitive to solvent, substituents, concentration

EXERCISE 114

Answers on page 208

Figure 62 (opposite) gives the NMR spectrum for ethanol. Study it and then attempt the following questions.

a What is responsible for the peak at zero chemical shift?

b Draw the displayed formula for ethanol and identify three types of hydrogen atoms (i.e. hydrogen atoms that are in different environments).

c Which hydrogen atoms in ethanol are responsible for the peaks with chemical shift (δ) between:

  i) 4 and 5,   ii) 3 and 4,

  iii) 1 and 2? [Hint: since these ranges are not in Table 19, you may need to decide which of the H atoms are most shielded.]

d What is the ratio of the peak areas and what is the significance of the value?

e Draw a sketch, with explanations, of the NMR spectrum you would expect from an isomer of ethanol called methoxymethane CH_3—O—CH_3.

Figure 62
NMR spectrum of ethanol.

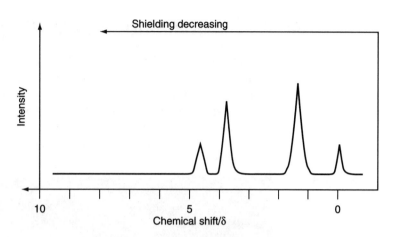

EXERCISE 115
Answers on page 209

Musk xylene is a synthetic perfume that has a pleasant musky smell. It is used widely in soaps and cosmetics. Its structure is very different from that of muscone, which is present in the natural musk obtained from the scent gland of the male musk deer, found in the mountains of Central Asia.

$$C(CH_3)_3$$

$$NO_2 \quad NO_2$$

$$CH_3 \quad CH_3$$

$$NO_2$$

musk xylene

Use chemical shift values given in Table 19 to predict what you would expect the NMR spectrum of musk xylene to look like.

Draw in the signals you would expect to see on a copy of the spectrum below, indicating their relative heights:

Figure 63

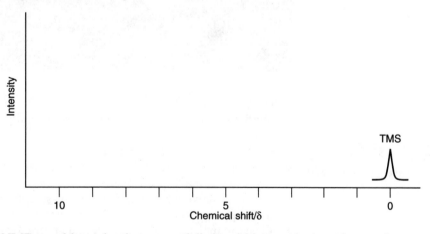

Thus NMR provides a simple means of distinguishing synthetic and natural compounds.

The chemical shift values and areas under the peaks, as we have seen, give important information about the number and type of hydrogen (protons) within a molecule. Further information that can be obtained from high resolution NMR spectrum is that derived from spin-spin splitting. Since very few syllabuses require you to know about this we have included it in Appendix 4. Check with your teacher whether you should study this as well.

The fact that NMR spectroscopy gives information about protons in different environments has led to some very interesting and important uses as you will now see.

■ 7.11 Applications of NMR spectroscopy

The application of NMR in medicine is becoming increasingly common and has become an important diagnostic tool in medicine in body scanners. Protons in water, proteins, lipids and most body sites give different signals in NMR spectra. This enables different organs of the body to be differentiated. Below is an NMR image of a human brain. This technique has no known side-effects and is now available in many clinics: it has been used to diagnose a variety of conditions such as cancer, hydrocephalus and multiple sclerosis.

NMR image of human brain.

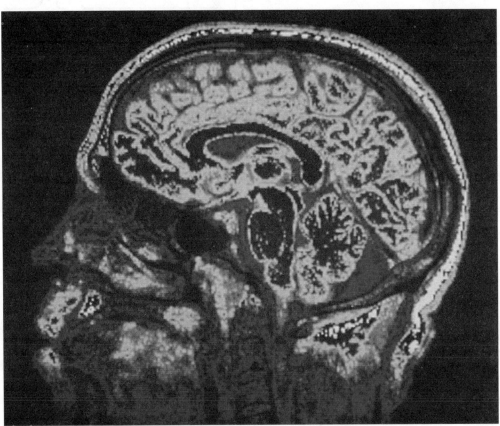

Before you go on to use all these types of spectra to identify organic compounds you summarise the contributions made by each technique.

EXERCISE 116
Answers on page 210

Each of the following statements describes the contribution made by a particular spectroscopic technique. Read each and identify it as either IR spectroscopy, NMR spectroscopy or mass spectrometry.
A. Gives a relative molecular mass for the sample, may give the number of carbon atoms present and may provide evidence for the presence of halogen atoms.
B. Gives an indication of the functional groups present in or absent from the sample.
C. Gives information concerning the structure of hydrogen-containing groups within the molecule.

Detailed examination of each spectrum in turn generally enables a final detailed structure to be determined. In the next section we give you a worked example on how this might be achieved in determining the identity of an unknown compound.

■ 7.12 Combined spectroscopic techniques

OBJECTIVE   When you have finished this section you should be able to:
■ use evidence from up to three spectra to suggest a probable structure for a given compound.

To give you an idea of how we use evidence from different spectra, to identify a compound, we show you a worked example. If one of the techniques is not a requirement of your particular syllabus, just concentrate on the other two.

WORKED EXAMPLE   Compound A has a composition by mass of carbon 54.5%, hydrogen 9.1%, and oxygen 36.4%. The IR spectrum, mass spectrum and NMR spectrum are shown below:

a Calculate the empirical formula of compound A.

b Determine the structure of compound A, name it and explain how the evidence leads to your conclusion.

Figure 64
Mass spectrum of A.

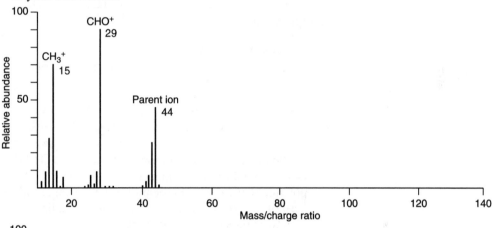

Figure 65
IR spectrum of A.

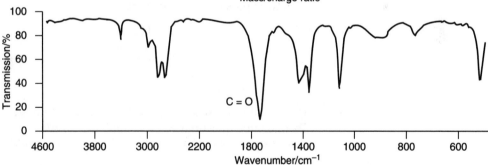

Figure 66
NMR spectrum of A.

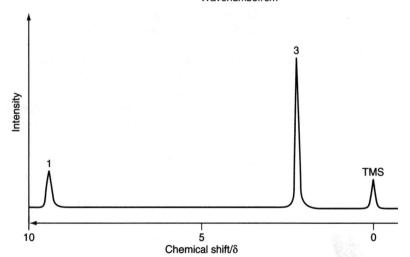

Solution

1. Calculate empirical formula from percentage composition by mass, using the method in previous exercises:

Table 20

	C	H	O
Mass/g	54.5	9.1	36.4
Amount/mol	$\dfrac{54.5}{12.0} = 4.54$	$\dfrac{9.1}{1.0} = 9.10$	$\dfrac{36.4}{16.0} = 2.28$
$\dfrac{\text{Amount}}{\text{Smallest amount}}$ = relative amount	$\dfrac{4.54}{2.28} = 1.99$	$\dfrac{9.1}{2.28} = 3.99$	$\dfrac{2.28}{2.28} = 1$
Simplest ratio	2	4	1

The empirical formula is C_2H_4O.

2. Calculate molecular formula from empirical formula from step 1 and molar mass from mass spectrum:

$$M_r \text{ of empirical formula} = (2 \times 12) + (4 \times 1) + (1 \times 16) = 44$$
$$\text{mass spectrum gives a molecular ion at } M_r = 44$$

So molecular formula = empirical formula = C_2H_4O.

3. Before looking at the evidence from spectra it is worth attempting to draw some possible structures from the molecular formula. This might give you some idea at least of the types of bonds and functional groups you might be looking for. In this case two* possible structures are:

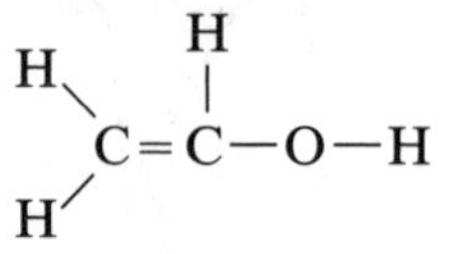

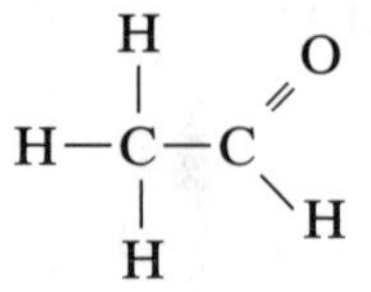

It may not be quite as simple as this for compounds containing more carbon atoms. Examination questions often include the results of chemical testing which at least gives you an idea of the functional groups you should be looking for in each type of spectrum.

a Infra-red spectrum

Look at the wavenumbers of the various dips (or troughs) and identify the bonds that are likely to be responsible, using the data in Fig. 45, page 114 to help you. Working from left to right, a dip at approximately 1700 cm^{-1} due to $C=O$ confirms A as a carbonyl compound.

b NMR spectrum

Shows two types of hydrogen atoms in the ratio 3:1. One hydrogen ($\delta \simeq 9$) is very unshielded; this is probably the one next to the carbonyl group and from Table 19

---

* You may have suggested a third possible structure, ethane oxide:

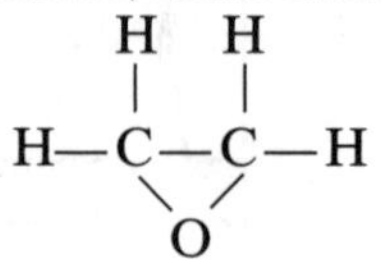

This is not on all A-level syllabuses, and exam boards would not expect it.

of chemical shift values (δ) on page 126, this is the type of proton from the group:

$$R-C\underset{}{\overset{\displaystyle\parallel O}{}}\!\!\!\diagdown_{\boxed{H}}$$

The three remaining hydrogens of $\delta = 2$, as we can also see from Table 19, must be those from the R group, in this case, CH_3.

c Mass spectrum

Scan along the spectrum and attempt to identify any peaks.

See if any of the peaks of mass/charge ratio add up to the mass/charge ratio to the parent ion, in this case, 44.

$$M_r = 29, \text{ corresponds to } CHO^+$$
$$M_r = 15, \text{ corresponds to } CH_3^+$$

In this example CHO^+ and CH_3^+ fragments must have formed when the C—C bond of the parent ion broke and ionised further.

Thus, the structure of compound A must be:

$$\underset{15}{\overset{\displaystyle H}{\underset{\displaystyle H}{H-C}}} - \overset{\displaystyle O}{\underset{\displaystyle H}{C}} \quad 29$$

ethanal

Not all analyses are as easy as this. Most examination questions include the results of chemical testing as well, which will give you a clue as to the types of bonds and functional groups you should be looking for. You then use the spectra to confirm what you suspect from chemical analysis.

 The next exercises give you practice with this type of question. You will need to consult Fig. 45, on page 114, and Table 19, on page 126, which give IR and NMR data respectively. In examinations this data will either be provided or is included in your data book which your examination board may require you to take into the examination. You may also wish to refer to Tables 21 and 22 (pages 111–113) in ILPAC 8, Functional Groups, which give characteristic tests for different functional groups.

EXERCISE 117
Answers on page 210

Compound B contains C, H and O only. It has a composition by mass of carbon 48.6%, hydrogen 8.1%, and oxygen 43.3%. The compound is acidic and effervesces on addition of sodium carbonate solution. The mass spectrum, IR spectrum and NMR spectrum, are shown below.

Use this information to identify compound B showing your reasoning.

Figure 67
Mass spectrum for compound B.

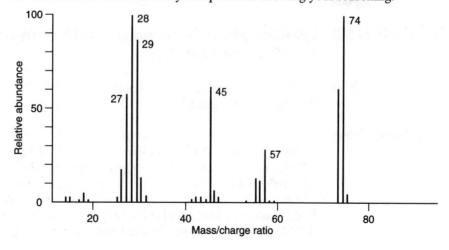

Figure 68
IR spectrum for
compound B.

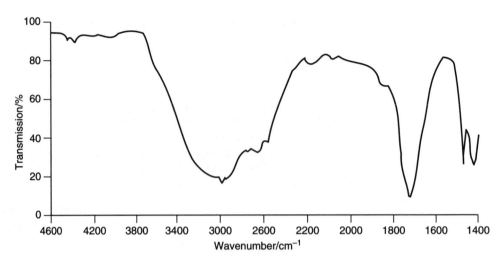

Figure 69
NMR spectrum for
compound B.

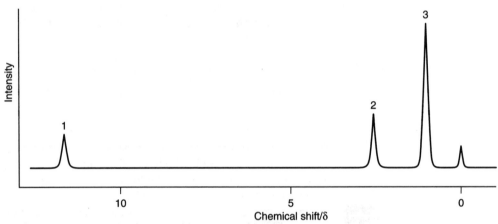

The next experiment requires you to identify three organic compounds from some
simple chemical tests together with elemental percentage composition, and the results of
mass and infra-red spectroscopy. Since the experimental work may be used for practical
assessment, specimen results and conclusions are not supplied in this book, but may be
obtained from your teacher. We have assumed that you may use your textbooks and
your own notes during the practical. You will find the table of characteristic tests for
functional groups in ILPAC 8, Functional Groups, page 111, Table 21 very useful.

If it is possible, and you have not already seen it, watch the second part of the ILPAC
video programme, 'Identifying Unknown Substances'. This will help you, not only with
the practical technique, but also with the interpretation of observations.

EXPERIMENT 8 Identification of three organic compounds from chemical testing and spectra

Aim To identify three organic compounds X, Y and Z which contain the elements carbon,
hydrogen and oxygen only.

Introduction In part (A) of the investigation you carry out some simple chemical tests specified in the
results table and record your observations and inferences (about the presence or
absence of certain functional groups) in (larger copies of) the tables provided. In part
(B) your teacher will give you percentage elemental analysis data together with mass
and IR spectra to not only confirm the functional groups suspected in part (A) but also
to determine the identity of each compound. You will be assessed only on part (A)
'Chemical testing' on your detailed observations and inferences.

Since you should preferably work under examination conditions, all the apparatus and chemicals should be provided for you so we have not included a requirements list. However, you must take a note of the following hazard warning.

HAZARD WARNING

  Compound Y and Z are highly flammable and the vapours are harmful. Therefore you **must:**
- **keep the compounds away from flames;**
- **perform the experiment in a fume cupboard;**
- **wear safety spectacles and gloves.**

  2,4-dinitrophenylhydrazine is toxic and 2M sulphuric acid is corrosive. Therefore you **must:**
- **wear safety spectacles and gloves;**
- **do boiling point determinations in the fume cupboard.**

Procedure – Part A

Chemical Testing

Results Table 6

Test			Observations	Inferences
1.	a	Burn a small amount of X in a combustion spoon		
	b	Shake a small amount of X in a test-tube with 5 cm³ of water. First try cold water then hot. (Use a hot water-bath)		
	c	To a small amount of X in a test-tube add 5 cm³ of 0.5 mol dm⁻³ sodium carbonate solution		
	d	Determine the melting point of X		
2.	a	Burn a small volume of Y in a combustion spoon		
	b	Shake a small amount (1–2 cm³) of Y with 5 cm³ of water in a test-tube. First try cold water then hot. (Use a hot water-bath)		
	c	To a few drops of Y in a test-tube, add about 5 cm³ of a mixture of aqueous potassium dichromate(VI) and 2M H₂SO₄ and warm in a hot water-bath		
	d	Determine the boiling point of Y in a fume cupboard		
3.	a	Burn a small volume of Z on a combustion spoon		
	b	Shake a small amount (1–2 cm³) of Z with 5 cm³ of water in a test-tube. First try cold then hot water		
	c	Add a few drops of Z to 5 cm³ of a solution of 2,4-dinitrophenyl-hydrazine (Brady's reagent) in a test-tube		
	d	Determine the boiling point of Z in a fume cupboard		

Complete the following table as far as possible.

Results Table 7

Compound	Functional group	Reasoning
X Y Z		

Procedure – Part B

Analysis of mass and IR spectra

Using the results of quantitative analysis, IR and mass spectra from your teacher together with the results from part (A), identify, with reasoning, the structures for compounds X, Y and Z. You will also need access to data on melting points and boiling points of organic compounds.

You should now attempt the following teacher-marked exercise.

EXERCISE

Teacher-marked

Mass spectrometry, infra-red and NMR spectroscopy are complementary techniques in chemical analysis. Choose any two of these techniques to explain:
a the physicochemical principles involved in these techniques,
b what information may be obtained from the spectra.

Appendix 4 gives details of information that can be obtained from high-resolution NMR spectroscopy, i.e. 'spin-spin splitting'. Check your syllabus to see if this is a requirement of your particular examination board.

Another important physical method for determining structure is X-ray crystallography. Since this technique relies upon the diffraction of X-rays by crystals, it can be used only on solids in crystalline form, and it therefore has limited application in organic chemistry. We have shown you some examples of the use of X-ray crystallography in ILPAC 3, Bonding and Structure. We illustrated another application, working out the shapes of some complex proteins, earlier in this book. In Appendix 5 we outline the theory and technique of X-ray crystallography. You should only study this if it is a requirement of your particular syllabus.

You have now completed the last ILPAC book on organic chemistry. Before you attempt the end-of-unit test we give you some guidance in revising your organic chemistry and some guidance in answering examination questions.

8

ORGANIC CHEMISTRY – TIPS ON ANSWERING EXAMINATION QUESTIONS

OBJECTIVE When you have finished this chapter you should be able to:
■ answer questions relating to any functional group studied in this book and in ILPAC 5, Introduction to Organic Chemistry, and in ILPAC 8, Functional Groups.

Some of the organic questions frequently set in examinations fall into the following categories:

■ synthetic pathways,
■ identification from chemical testing and/or spectroscopic techniques,
■ distinguishing tests,
■ predicting chemical reactions from a given structure,
■ writing mechanisms.

You have met all of these types of organic questions in all three organic books; now we bring them together with some useful tips on how to revise and tackle each type.
    We give you an example of each type of question and a few tips on what you should revise before tackling it. Make sure you read the tips before tackling each question.
    We also show how marks may be allocated in the answers.

■ 8.1 Synthetic pathways

EXERCISE 118
Answers on page 211

Outline the reactions you would carry out to complete each of the syntheses shown below in not more than three stages.
    For each reaction, state the reagents you would use and the conditions under which the reaction would occur. (You are **not** expected to describe how any of the products are purified).

a

$$H-\underset{\underset{H}{|}}{\overset{\overset{H}{|}}{C}}-\underset{\underset{H}{|}}{\overset{\overset{H}{|}}{C}}-O-H \longrightarrow H-\underset{\underset{H}{|}}{\overset{\overset{H}{|}}{C}}-\underset{\underset{H}{|}}{\overset{\overset{H}{|}}{C}}-C\overset{\overset{\displaystyle O}{\diagup}}{\underset{\displaystyle O-H}{\diagdown}}$$

ethanol   propanoic acid

b

$$Cl-\underset{\underset{H}{|}}{\overset{\overset{H}{|}}{C}}-\underset{\underset{H}{|}}{\overset{\overset{H}{|}}{C}}-O-H \longrightarrow \underset{H}{\overset{H}{\diagdown}}N-\underset{\underset{H}{|}}{\overset{\overset{H}{|}}{C}}-C\overset{\overset{\displaystyle O}{\diagup}}{\underset{\displaystyle O-H}{\diagdown}}$$

2-chloroethanol   aminoethanoic acid

Read the notes below before tackling this.   (18)

Tips

You must first make sure you revise all the chemical reactions from your summary prepared from the teacher-marked exercise on page 39. Also refer back to ILPAC 5, Introduction to Organic Chemistry, Section 7.10, where we give you a worked example in tackling this type of question. Remember, it is probably easier to work backwards from the target compound, i.e. mentally list the compounds that you know can be converted into the final target, then see if you can find a link between them and the starting compound. Go through some of the questions that appear near the end of each functional group, under the heading 'Synthetic pathways'. You will find page references to these on the contents page at the front of each book. Now you are ready to tackle the problem above.

■ 8.2 Identification of compounds from chemical testing and/or spectroscopic techniques

EXERCISE 119

Answers on page 212

The following reactions were observed for a compound G of formula C_3H_6O.
1. The compound did not react with alkaline aqueous copper(II) ions, even when heated.
2. On adding 2,4-dinitrophenylhydrazine, a yellow–orange precipitate formed.
3. Reaction with hydrogen in the presence of a catalyst produced a colourless liquid H. Liquid H reacted with sodium to give hydrogen.
   a Draw the displayed (full structural) formulae of two compounds of formula C_3H_6O.
   b What does the result of reaction 1 show?
   c The formation of a yellow–orange precipitate in reaction 2 is a positive test for a particular organic group. Identify this group.
   d Using the formula of the compound and the results of reactions 1 and 2, identify G.

Read the tips below before tackling this exercise. (5)

EXERCISE 120

Answers on page 212

Analysis of an organic compound Z containing carbon, hydrogen and oxygen gave the following data:

> composition, by mass: C, 66.7%; H, 11.1%; O, 22.2%
> infra-red spectrum: strong absorption band at 1715 cm^{-1}
> mass spectrum: lines of m/e values 72, 57, 43

Use these data to suggest the identity and molecular structure of Z, showing your reasoning. (10)

Read the tips below before tackling this exercise.

Tips

For questions like Exercise 119, which give results of chemical testing, you should learn the characteristic tests for all the different functional groups from ILPAC 8, Functional Groups, Tables 21 and 22 (pages 111–113). For questions that also include spectra, as shown in Exercise 120, you should go over the relevant section on 'Physical methods for determining structure', pages 109 to 132 in this book. When you have done all this you can tackle Exercises 119 and 120.

Sometimes you may have to work out the identity of a series of compounds involved in a synthetic pathway from the results of chemical testing. Remember in this type of question you may find it useful to summarise the information in a series of steps first. Have another look at the method shown in the answer to Exercise 100 and then tackle Exercise 121 below.

EXERCISE 121

Answer on page 212

A compound **D**, C_7H_6O, reacts with 2,4-dinitrophenylhydrazine to form an orange–yellow precipitate. **D** is oxidised by acidified aqueous potassium dichromate(VI) to form the compound **E**. The reducing agent lithium aluminium hydride converts **D** to compound **F**. On treatment with aqueous sodium hydroxide, **E** gives compound **G**.
a Deduce the identity of compounds **D**, **E**, **F** and **G**.
b Explain each of the reactions indicated below.
   i)   **D** + 2,4-dinitrophenylhydrazine
   ii) **D** + acidified aqueous potassium dichromate(VI)
   iii) **D** + lithium aluminium hydride
c What type of reaction would **D** undergo with hydrogen cyanide?
d i) Draw the displayed formula of the product of reaction in (c).
   ii) What form of isomerism does this compound show?

■ 8.3 Distinguishing tests

EXERCISE 122

Answers on page 213

For each of the following pairs of compounds, describe one simple chemical test that would enable you to distinguish between them. In each case, state clearly how each member of the pair behaves in the test and explain the reaction involved, writing balanced equations where appropriate.

a
(benzene ring) and (benzene ring with $CH=CH_2$)

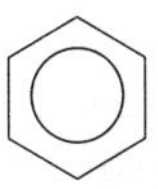

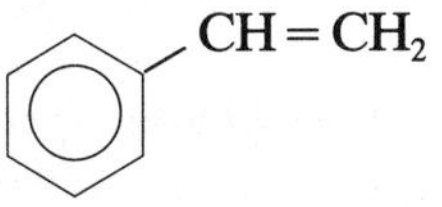

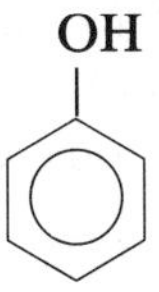

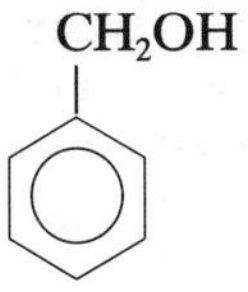

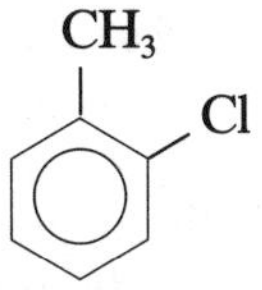

b
(benzene ring with OH) and (benzene ring with CH_2OH)

c
(CH_3, Cl on benzene ring) and (benzene ring with CH_2Cl)

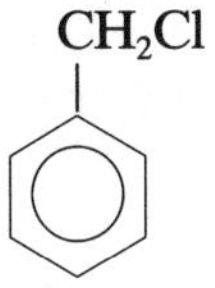

Read the notes below before attempting this exercise.   (9)

Tips

The simple chemical test you choose in this type of question should give an **observable** difference such as those listed in Table 21 'Characteristic tests for organic compounds', on pages 111–113 of ILPAC 8, Functional Groups. You should also refer to the questions we give under the heading 'Distinguishing tests' in each organic book. The contents lists should help you find these. Don't forget that if one substance of a pair does not react in a particular test, you must **state** this – it is not enough to give only the positive result. When you have done all this, tackle the question above.

■ 8.4 Predicting chemical reactions from a given structure

EXERCISE 123

Answers on page 213

Predict how the compound whose structure is shown below will react with:
a bromine (in inert solvent),
b phosphorus pentachloride,
c lithium tetrahydridoaluminate(III).

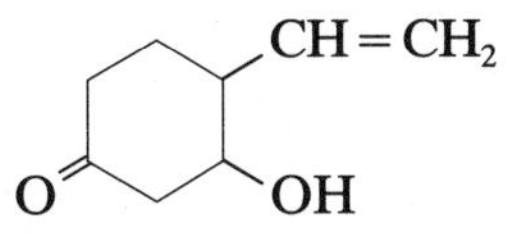

(ring structure with $CH=CH_2$, O, OH)   (6)

Write the structure of the product in each case. Read the notes below before attempting this question.

Tips

The key to answering this type of question is to realise that the chemistry of carbon compounds is mainly the chemistry of functional groups. You can see in the structure above a double bond, as in an alkene; an —OH group like that of an alcohol (not like that of a phenol since the ring is saturated), and a $C=O$ group like that of a ketone. If you know the chemistry of these groups you can predict how they will react with the reagents specified. Once again you need to know all your reactions from your summary chart. Now tackle the question above.

■ 8.5 Writing mechanisms

EXERCISE 124

Answers on page 214

What do you understand by the terms:

a electrophilic substitution,

b nucleophilic substitution?

Illustrate your answers by describing the mechanisms of **one** example of each type of reaction. Before attempting this read the notes below. (10)

Tips Your particular syllabus specifies which particular reactions you should know the mechanisms for. You could copy them all down on a single side of A4. Make sure you are confident with the terminology used and do not get slapdash with curly arrows. They must be precisely positioned. Sections 2.6 and 3.8 of ILPAC 5, Introduction to Organic Chemistry, should help you with this. When you are quite sure you know all the mechanisms tackle the question above.

These are just a selection of some of the types of questions you may be set in your examinations. Remember when you are preparing your revision notes that your syllabus is your best guide.

The end-of-unit test that follows will include questions which not only test your knowledge of this book but the other organic books as well. You should therefore make sure you revise **all** your organic books thoroughly before attempting it. Good luck!

■ End-of-unit test

To find out how well you have learned the material in this book try the test which follows. Read the notes below before starting.
1. You should spend about 120 minutes on this test.
2. Hand your answers to your teacher for marking.

Part A
Answer **all** the questions in this section
1. The table below gives data about a number of amino acids which occur in proteins.

Table 21

Name and abbreviation	Relative molecular mass	R_f value in solvent I	R_f value in solvent II
Alanine, ala	89	0.43	0.38
Aspartic acid, asp	133	0.13	0.24
Glycine, gly	75	0.33	0.26
Leucine, leu	131	0.66	0.73
Lysine, lys	146	0.62	0.14
Phenylalanine, phe	165	0.64	0.68
Serine, ser	105	0.30	0.27
Valine, val	117	0.58	0.40

A small polypeptide was hydrolysed with concentrated acid and, after neutralisation, the resulting amino acids were separated by two-way chromatography. The chromatogram is shown below.

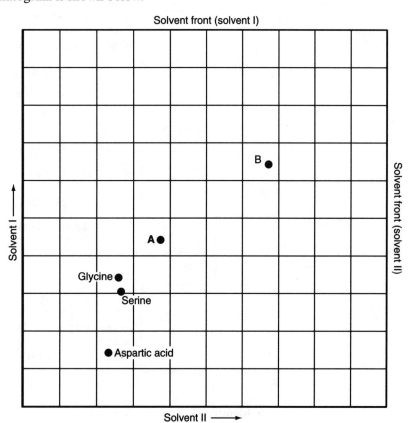

The R_f value is given by the distance travelled by an amino acid spot divided by the distance travelled by the solvent front.

a Determine the R_f values in both solvents of the amino acids labelled A and B on the chromatogram and, hence, identify them.

Table 22

Amino acid	R_f value in solvent I	R_f value in solvent II	Identity
A			
B			

(4)

b Indicate by grid reference where you would find the spot on the chromatogram corresponding to lysine. (1)

c By reference to the table of data, explain why two-way chromatography is needed to separate these eight amino acids. (2)

d Explain why the R_f values of an amino acid differ in different solvents. (1)

e When the solvents have moved up the chromatography paper, how is the paper treated to locate the positions of the amino acids? (2)

f Why is it important to avoid fingering the chromatography paper during the experiment? (1)

2. **a** Complete the reaction scheme shown below which starts with the compound ethene. Write the structural formula of the principal organic product or intermediate compound for compounds A to F.

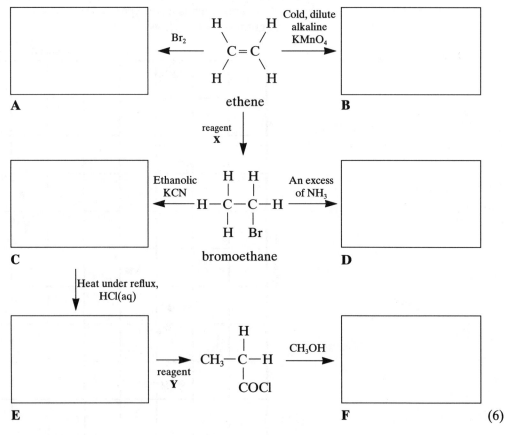

b Identify the reagents X and Y (2)

3. The manufacture of the polymer commonly known as polystyrene starts from the production of the monomer in a two-stage process.

Stage 1 is a Friedel–Crafts reaction between ethene and benzene to form ethylbenzene:

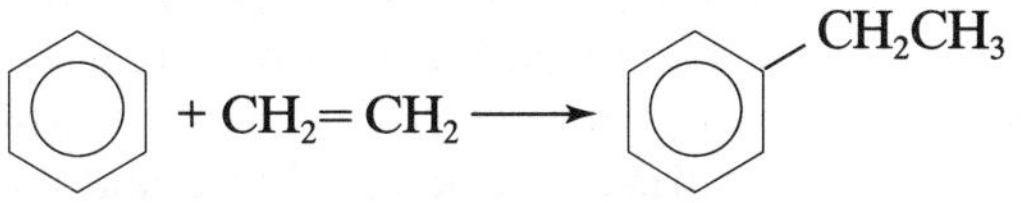

The ethylbenzene is separated from other products of side-reactions before use in stage 2.

Stage 2 is the catalytic dehydrogenation of ethylbenzene to styrene monomer (molecular formula C_8H_8), which is then separated and polymerised.

a Suggest a suitable catalyst for the Friedel–Crafts reaction to form ethylbenzene in stage 1. (1)

b i) What is meant by catalytic dehydrogenation?

ii) Hence write down the equation for the reaction in stage 2, giving the structural formula for styrene. (3)

c Ethylbenzene reacts slowly when liquid bromine is added in the presence of a catalyst. However, styrene reacts rapidly with bromine even in dilute aqueous solution.

Explain these differing reactivities of the two hydrocarbons with the same reagent. (4)

d The reaction in stage 2 is endothermic. It is carried out in the gas phase, and results in an equilibrium mixture of reactant and products.

Describe the conditions of temperature and pressure that are likely to be used in order to maximise the yield of styrene at equilibrium. (2)

e i) Draw a diagram to illustrate the structure of a polystyrene molecule, including at least two monomer units.

ii) Name the type of polymerisation that has taken place.

iii) Polystyrene is noted for being a tough plastic material, with limited flexibility.

Use your knowledge of polymer structures and your answer to (i) to suggest why polystyrene has these properties. (4)

4. The following scheme indicates a sequence of single-stage reactions to prepare aspirin from methylbenzene:

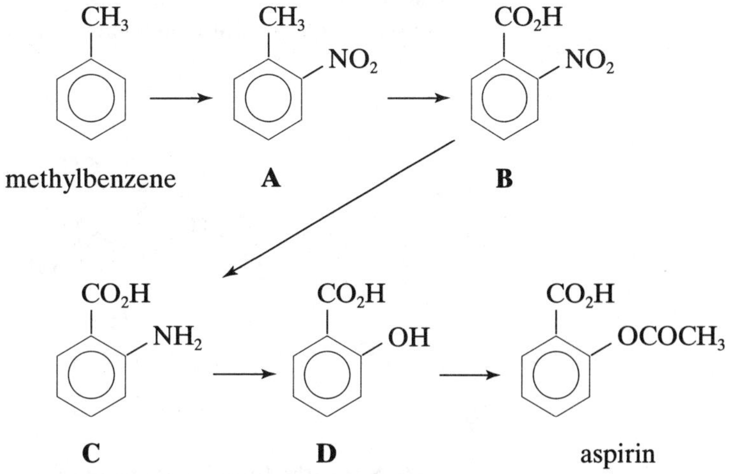

a Give the reagents and conditions required to convert:

i) methylbenzene into A,

ii) A into B. (3)

b D is converted into aspirin by an esterification reaction.
  i) Name the reagents and suggest appropriate conditions for the esterification.
  ii) During esterification, traces of a polymeric by-product always form. Explain why it is formed and suggest a structure for it. (6)
c i) A is an isomer of C. The type of isomerism exhibited can be described as one form of structural isomerism. Which form?
  ii) Suggest, giving a reason, which of A and C would have the higher melting point.
  iii) Give a chemical test, and the expected result, which would enable A and C to be distinguished. (7)
d 4.60 g of methylbenzene gave 0.90 g of aspirin.
  i) Calculate the percentage yield obtained.
  ii) What feature of the proposed synthesis is mainly responsible for the low yield? (3)

5. The compound A (pentyl 4-methoxycinnamate) is used in some sunscreen creams to protect the skin from burning in ultraviolet rays from the sun. The formula of A is:

$$CH_3-O-\langle\bigcirc\rangle-CH=CH-C\begin{smallmatrix}\nearrow O\\\searrow OCH_2(CH_2)_3CH_3\end{smallmatrix}$$

a i) Name three functional groups present in A.
  ii) Describe a test to show the presence of unsaturation in this molecule. Give reagents, conditions and an equation for this reaction as well as the observable result. (6)
b Give the structures of the organic products formed when A undergoes the following reactions:
  i) Heating with hydrogen in the presence of a nickel catalyst.
  ii) Heating under reflux with aqueous sodium hydroxide. (3)
c Two isomeric forms of the compound A exist. Give the structures of these two isomers, state the type of isomerism and explain briefly how it occurs. (4)
d Explain briefly what environmental changes are occurring which may require an increase in the use of such sunscreen creams in the future. (3)

6. Oil of peppermint is a plant extract obtained from the leaves and stems of *Mentha piperita*. It is a complex mixture of organic compounds, which is used widely in perfumery, food flavouring, toothpastes and a range of pharmaceutical products. The main component (which contributes about 46% by mass of oil of peppermint) is menthol, which can be separated out as crystals when the oil is cooled. The crystals of menthol are only slightly soluble in water, but they are moderately soluble in warm ethanol.

$$
\begin{array}{c}
CH_3\\
|\\
CH\\
CH_2 \diagup \quad \diagdown CH_2\\
|\qquad\qquad |\\
CH_2 \diagdown \quad \diagup CH\\
CH \quad OH\\
|\\
CH\\
CH_3 \diagup \quad \diagdown CH_3
\end{array}
$$

(Melting point = 42.5°C, M_r = 156)

menthol

a i) Name the functional group present in menthol. (1)
  ii) Classify this functional group as primary, secondary or tertiary. (1)
  iii) The infra-red (IR) spectrum of a sample of Brazilian oil of peppermint is shown below.

Infra-red spectrum of a
sample of Brazilian oil of
peppermint.

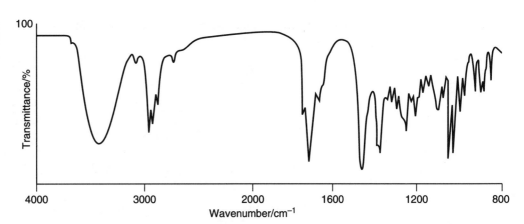

Write down the wavenumber of an absorption in the IR spectrum which is
characteristic of the functional group present in menthol. (Refer to the table of
characteristic IR absorptions, page 114.)   (1)

b Another substance, compound X, can be separated as a colourless liquid from oil
of peppermint. Its IR spectrum and mass spectrum are shown in the two diagrams
below. Compound X can be made from menthol in a simple one-step process.

Infra-red spectrum of
compound X.

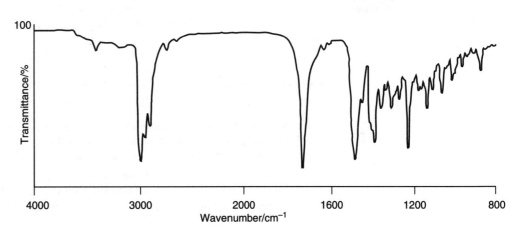

Six most significant peaks in
the mass spectrum of
compound X.

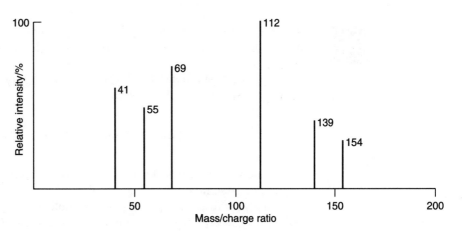

i) Use the information contained in these two diagrams to suggest a structure for
compound X and explain your reasoning.   (5)

ii) What reagent(s) and conditions would you use to convert menthol into
compound X?   (2)

c Synthetic menthol can be made by the following route:

i) Give the reagents and conditions which could be used to carry out step A. (3)

ii) What name is given to the reaction in step A? (1)

iii) A simple test involving the OH group can be used to determine when step B has gone to completion. Describe the test and how you would carry it out. (3)

Part B
Answer only **one** question from this section.

1. The amino acid L-tyrosine has the formula:

$$HO-\langle\bigcirc\rangle-CH_2-\underset{\underset{NH_2}{|}}{CH}-CO_2H$$

Use your knowledge of organic chemistry to describe the major physical and chemical properties you would expect L-tyrosine to have.

For any reactions that you describe, give appropriate details of the reagents, conditions and structural formulae of the products.

As a guide, an answer that gives a sound and balanced account of predictions for at least two physical properties and eight chemical reactions, explaining the way in which the main features of the L-tyrosine molecule give rise to these properties, with accompanying equations and conditions, is likely to be awarded a high mark. (15)

2. When preparing esters in the laboratory, moderate yields may be obtained by reaction of an alcohol with a carboxylic acid. The reaction is usually slow and incomplete.

Plan a reaction scheme for the preparation of butyl ethanoate using this reaction, starting from butan-1-ol and ethanoic acid.

carboxylic acid + alcohol → ester + water

Calculate the quantities of starting materials required to produce 10 g of butyl ethanoate, assuming a 50% yield.

Using your result, describe how you would carry out this one-stage synthesis in the laboratory, in order to produce a pure sample of about 10 g of butyl ethanoate. Give sufficient details of apparatus, procedures and safety precautions to enable a fellow A-level student to perform the experiment.

What will you need to do to speed up the rate of reaction? What can be done to counter the incomplete nature of the reaction? Table 23 and the information that follows it should help you write your answer.

Table 23

	Molar mass /g mol^{-1}	Density /g cm^{-3}	Boiling point /°C
Ethanoic acid	60.1	1.049	118.0
Butan-1-ol	74.1	0.810	117.3
Butyl ethanoate	116		126

(15)

!Hazards
Butan-1-ol is **flammable and harmful**. It is harmful by inhalation: irritating to eyes and skin; readily absorbed through the skin.

Ethanoic acid is **flammable and corrosive**. It causes severe burns; extremely irritant to all tissues.

(Total: 100 marks)

AN ALTERNATIVE ESTER PREPARATION

This experiment is an alternative to Experiment 3 Preparation of aspirin, page 31. Since your teacher may use this for practical assessment we do not give specimen results.

EXPERIMENT 9   Preparing phenyl benzoate

Aim   The purpose of this experiment is to prepare a sample of phenyl benzoate, purify it by recrystallisation, and measure its melting point.

Introduction   You prepare phenyl benzoate by shaking phenol with benzoyl chloride in alkaline solution:

$$C_6H_5COCl + C_6H_5OH \rightarrow C_6H_5CO_2C_6H_5 + HCl$$

The product appears as a solid and you purify it by filtering, dissolving in hot ethanol, and cooling to recrystallise the ester. This method is known as the Schotten–Baumann reaction.

Requirements – Part A
- safety spectacles and protective gloves
- weighing bottle
- spatula
- phenol, C_6H_5OH
- access to balance, sensitivity ± 0.01 g
- conical flask, 250 cm^3, with tight-fitting rubber bung
- measuring cylinder, 100 cm^3
- sodium hydroxide solution, 2 M NaOH
- measuring cylinder, 10 cm^3
- benzoyl chloride, C_6H_5COCl
- suction filtration apparatus (see Fig. 70)
- wash-bottle of distilled water

– Part B
- boiling-tube
- ethanol, C_2H_5OH
- glass rod
- water-bath or 250 cm^3 beaker
- Bunsen burner, tripod, gauze and bench mat
- thermometer, 0–100 °C
- ice
- suction filtration apparatus (see Fig. 70)
- filter papers
- specimen bottles
- access to balance, sensitivity ± 0.01 g

– Part C
- melting point tubes (at least 2)
- watch-glass
- thermometer, 0–100 °C, long stem
- boiling-tube fitted with cork and stirrer
- rubber ring or band
- dibutyl benzene-1,2-dicarboxylate (dibutyl phthalate)
- retort stand, boss and clamp

     } or electrical melting point apparatus

HAZARD WARNING

  Benzoyl chloride is lachrymatory (produces tears), is irritating to the skin and can cause burns.

 Phenol vapour is harmful to the eyes, lungs, and skin. Solid and solution are corrosive and poisonous by skin absorption.

 Sodium hydroxide solution is very corrosive. Even when dilute it can damage your eyes.

 Ethanol is flammable.
Therefore you **must**:
■ **work at a fume cupboard;**
■ **wear safety spectacles and protective gloves;**
■ **keep the stoppers on bottles;**
■ **keep ethanol away from naked flames.**

Procedure – Part A

Preparation of phenyl benzoate

1. Transfer about 5.0 g of phenol into a weighing bottle and weigh it to the nearest 0.01 g.
2. Into a conical flask pour 90 cm^3 of 2 M sodium hydroxide and the bulk of the phenol from the weighing bottle.
3. Reweigh the weighing bottle, with any remaining phenol, to the nearest 0.01 g.
4. In a fume cupboard pour 9 cm^3 of benzoyl chloride into the conical flask.
5. Insert the bung securely and shake the bottle for 15 minutes, carefully releasing the pressure every few minutes as the flask gets warm.
6. Cool the flask under cold, running tap water.
7. Filter the crude product using a suction filtration apparatus (Fig. 6(b), page 33). Use a spatula to break up the lumps of ester on the filter paper, being careful not to puncture the filter paper.
8. Pour more water over the crude ester to destroy any remaining benzoyl chloride.

– Part B

Recrystallisation

1. Transfer the crystals to a boiling-tube and just cover them with ethanol.
2. Place the boiling-tube in a water-bath or beaker of hot water, kept at about 60 °C, and stir with a glass rod.
3. If some solid ester is still visible, add **just** enough ethanol to dissolve it completely after stirring.
4. In order to allow the separation of the ester as a solid rather than an oily liquid (phenyl benzoate has a low melting point) add more ethanol to double the volume of solution.
5. Cool the solution in an ice–water mixture until crystals appear.
6. Filter the crystals through the suction apparatus, using a clean Büchner funnel and filter paper. To avoid losing any solid, break the vacuum and use the filtrate to rinse the boiling-tube into the funnel.
7. Using suction again, rinse the crystals with about 1 cm^3 of **cold** ethanol and drain thoroughly.
8. Press the crystals between two wads of filter paper to remove excess solvent. Then put the crystals on another dry piece of filter paper placed alongside a Bunsen burner and gauze, turning the crystals over occasionally until they appear dry. Don't let them get too hot or they will melt!

9. Weigh the dry crystals in a pre-weighed specimen bottle and record the mass of your sample of phenyl benzoate. Calculate the percentage yield using the method described in ILPAC 5, Introduction to Organic Chemistry, page 95 and the expression:

$$\% \text{ yield} = \frac{\text{actual mass of product}}{\text{maximum mass of product}} \times 100$$

– Part C **Determination of melting point.**

Follow the procedure that is described in Experiment 3 (page 34) and record the melting point of your sample of phenyl benzoate.

Hand in your product suitably labelled, i.e. mass, melting point, name and date.

Results Table 8

Mass of weighing bottle and phenol	g
Mass of weighing bottle after emptying	g
Mass of phenol	g
Mass of specimen bottle	g
Mass of specimen bottle and phenyl benzoate	g
Mass of phenyl benzoate	g
Melting point of phenyl benzoate	°C

APPENDIX 2

NUCLEIC ACIDS (DNA AND RNA)

These remarkable substances, found both inside and outside the cell nucleus, are capable of reproducing themselves and of storing the information needed to assemble hundreds of amino acids in the correct order for making protein molecules. In this section you examine the chemical composition of deoxyribonucleic acid (DNA) and ribonucleic acid (RNA) and see how their structures fit them for their biological function.

OBJECTIVES

When you have finished this appendix you should be able to:

- name the three components of a **nucleotide**;
- show how **polynucleotides** are formed from nucleotides;
- identify **purine** and **pyrimidine** bases from their formulae and show how one base of each type can be linked by hydrogen bonds.

Nucleic acids, like proteins, consist of long chains containing a great many sub-units. However, whereas a protein may have twenty **different** sub-units (amino acids), nucleic acids usually have four, called nucleotides. Because of this, nucleic acids are often called polynucleotides.

Each nucleotide consists of three parts:

1. A monosaccharide (usually ribose or deoxyribose).
2. Phosphoric(V) acid (esterified with an —OH group in the sugar).
3. A base (attached by condensation with another —OH group in the sugar).

In a polynucleotide, each phosphoric acid molecule becomes esterified with two sugar units, forming what is conveniently called the 'sugar-phosphate' backbone, to which four different bases are attached at regular intervals.

It is the order in which these four bases are attached that distinguishes the DNA molecules of different species and encodes all the genetic information necessary for cell replication and protein synthesis.

Read about nucleic acids in a textbook of organic chemistry or simple biochemistry. Look for the difference in the sugar/phosphate backbone between DNA and RNA. Also notice which purine and pyrimidine bases are attached to the sugar/phosphate backbone and the way in which hydrogen bonding holds pairs of chains together. You need not attempt to learn the complete structures but should find enough information to be able to do the exercise that follows.

EXERCISE 125

Answers on page 214

Figure 71 shows the sugar phosphate backbones of two DNA chains with attached purine and pyrimidine bases.

Figure 71

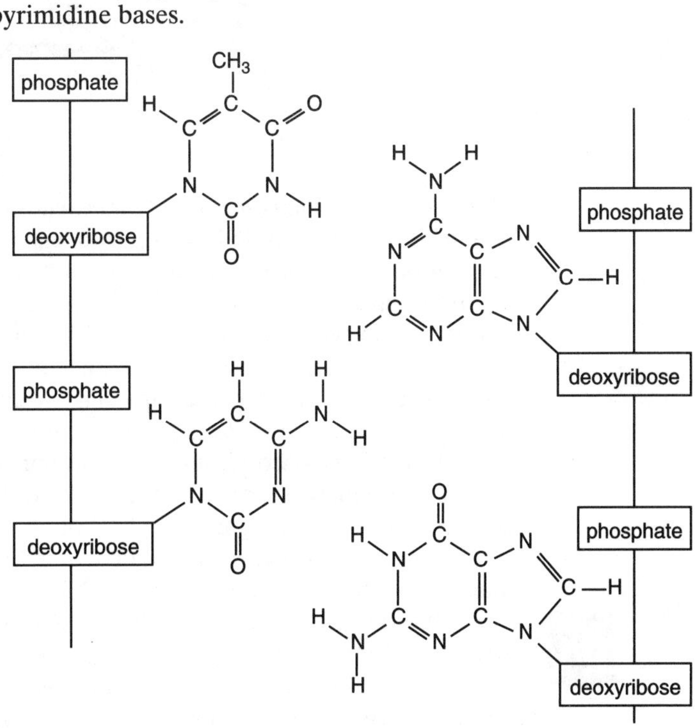

a Name the purine bases and pyrimidine bases.
b On a copy of Fig. 71 draw in the sites where you would expect hydrogen bonding to take place.
c State two ways in which an RNA polynucleotide differs in structure from a DNA polynucleotide.

As you saw in the exercise you have just completed, the two strands of DNA are joined by their pairs of complementary bases. Normally, the chains are twisted into the famous double helix, shown as a diagram in Fig. 72.

Figure 72
Replication of the DNA double helix.

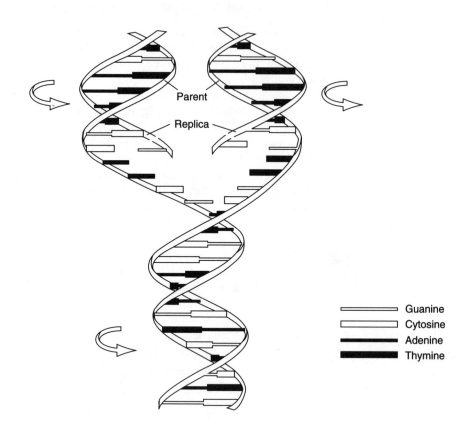

However, if the chains separate from each other, as shown in Fig. 72, the base pairing enables a replica chain to build up from each parent chain. This is the basis of DNA replication which takes place during cell division or whenever a new protein is to be synthesised.

For protein synthesis, however, an RNA molecule builds up on part of the parent DNA strand and moves outside the nucleus to the ribosomes. Here, it acts as a template. Every group of three bases (a triplet) is the code for a particular amino acid so that the appropriate amino acids for a particular protein gradually assemble in the right order.

Each step of DNA replication and protein synthesis is catalysed by a specific **enzyme**, another protein, which enables the reaction to take place at the temperature and under the conditions prevailing in the cell. Protein synthesis and DNA replication are fascinating topics which we cannot cover in detail in this course but which you would learn more about in a more advanced biochemistry or cell biology course.

For background information about the way in which scientists worked out the structure of DNA, you should read *The Double Helix* by J D Watson. The book also gives a fascinating insight into the way scientific work is done and the rivalries (friendly and otherwise) between separate groups working on the same problem.

EMPIRICAL FORMULA FROM ELEMENTAL ANALYSIS DATA

On page 110 we showed you how to work out an empirical formula from mass percentage data. We now show you how it is determined from the results of elemental analysis. You need only do this if it is specified on your syllabus. First, we do a worked example for a compound containing carbon and hydrogen only.

WORKED EXAMPLE

A 1.00 g sample of hydrocarbon was burnt in excess of oxygen. The carbon dioxide and water produced was absorbed in previously weighed tubes containing magnesium chlorate(VII) and soda-lime respectively. The tubes were found to have gained 3.38 g of carbon dioxide and 0.692 g of water. Determine the empirical formula of the compound.

Solution

1. Calculate the amount of carbon in the sample from the mass of carbon dioxide absorbed.

$$\text{amount of } CO_2 = \frac{\text{mass}}{\text{molar mass}} = \frac{3.38 \text{ g}}{44.0 \text{ g mol}^{-1}} = 0.0768 \text{ mol}$$

In 1 mol of CO_2 there is 1 mol of C,
∴ amount of C = amount of CO_2 = 0.0768 mol

2. Calculate the amount of hydrogen in the sample from the mass of water absorbed.

$$\text{amount of } H_2O = \frac{\text{mass}}{\text{molar mass}} = \frac{0.692 \text{ g}}{18.0 \text{ g mol}^{-1}} = 0.0384 \text{ mol}$$

In 1 mol of H_2O there are 2 mol of H atoms
∴ amount of H = 2 × amount of H_2O = 0.0768 mol

3. Calculate the empirical formula from the ratio C : H. In this case, the amounts of C and H are equal. Therefore, the empirical formula is C_1H_1, i.e. **CH**.

When oxygen is present, as well as carbon and hydrogen, the amount must be determined by difference. We now show this method.

WORKED EXAMPLE

A 1.00 g sample of compound A was burnt in excess oxygen producing 2.52 g of CO_2 and 0.433 g of H_2O. Calculate the empirical formula of the compound.

Solution

As before, determine the amounts of carbon and hydrogen atoms.

1. Amount of $CO_2 = \frac{\text{mass}}{\text{molar mass}} = \frac{2.52 \text{ g}}{44.0 \text{ g mol}^{-1}} = 0.0573 \text{ mol}$

∴ amount of C = 0.0573 mol

2. Amount of $H_2O = \frac{\text{mass}}{\text{molar mass}} = \frac{0.443 \text{ g}}{18.0 \text{ g mol}^{-1}} = 0.0246 \text{ mol}$

∴ amount of H = 2 × 0.0246 mol = 0.0492 mol

3. Now calculate the mass of carbon and mass of hydrogen.
   Mass of C = amount × molar mass = 0.0573 mol × 12.0 g mol^{-1} = 0.688 g
   Mass of H = amount × molar mass = 0.0492 mol × 1.00 g mol^{-1} = 0.0492 g
4. The mass of oxygen = mass of sample − (mass of C + mass of H)
   = 1.00 g − (0.688 g + 0.0492 g) = 0.263 g
5. Now construct a table:

Table 24

	C	H	O
Mass/g			0.263
Molar mass/g mol^{-1}			16.0
Amount/mol	0.0573	0.0492	0.0164
$\dfrac{\text{Amount}}{\text{Smallest amount}}$	$\dfrac{0.0573}{0.0164} = 3.49$	$\dfrac{0.0492}{0.0164} = 3.00$	$\dfrac{0.0164}{0.0164} = 1.00$
Simplest ratio of relative amounts	7	6	2

Empirical formula = $\mathbf{C_7H_6O_2}$.
Use similar methods in the next two exercises.

EXERCISE 126
Answer on page 215

Complete combustion of a hydrocarbon, Z, gives 0.66 g of carbon dioxide and 0.36 g of water. (Relative atomic masses: H = 1.0, C = 12, O = 16.)
What is the empirical formula of Z?

EXERCISE 127
Answer on page 215

The combustion of 0.146 g of compound B gave 0.374 g of carbon dioxide and 0.154 g of water. Assuming that B contains carbon, hydrogen and oxygen only, determine its empirical formula.

As you may have found out from your reading, one of the ways of determining the halogen content of a compound is to make from it a precipitate of silver halide, which is filtered off, washed, dried and weighed. The next exercise includes information that will enable you to determine the empirical formula of a halogeno-compound.

EXERCISE 128
Answer on page 216

0.2243 g of compound C gave 0.3771 g of carbon dioxide, 0.0643 g of water and 0.2685 g of silver bromide. Determine the empirical formula of the compound.

SPIN-SPIN SPLITTING IN NMR SPECTRA

This appendix is an extension to Section 7.10 which dealt with NMR spectra. The chemical shift values and areas of the peaks from NMR spectra gave important information about the number and type of protons contained within a molecule. Further information can be obtained from high-resolution spectra where the magnetic field is more accurately controlled. When this is done the single peaks due to the same types of hydrogen atoms can be split into different peaks. This is called spin-spin splitting.

OBJECTIVES

When you have finished this appendix you should be able to:
■ predict, from an NMR spectrum, the number of protons adjacent to a given proton using **spin-spin splitting** as a diagnostic tool;
■ describe how the addition of $^2_1H_2O$ (D_2O) may be used to identify **labile protons**.

Read about spin-spin splitting from a suitable source suggested by your teacher. Since very few chemistry textbooks cover this we lead you through the next few exercises with rather more guidance and information than usual.

EXERCISE 129
Answers on page 216

Consider the molecule 1,1,2-trichloroethane shown below:

$$\begin{array}{ccc} H_a & H_b & \\ | & | & \\ Cl-C-C-Cl \\ | & | & \\ Cl & H_b & \end{array}$$

As you can see there are two chemically distinct types of proton, H_a and H_b. On this basis predict:
a the number of resonance peaks on its low resolution NMR spectrum,
b the ratio of the areas under the peaks.

At high-resolution, the NMR spectrum is a little more complex as shown below. The single proton, H_a, produced the triplet of peaks and the two protons, H_b, produced the doublet of peaks.

Figure 73
High-resolution NMR spectrum for 1,2-trichloroethane.

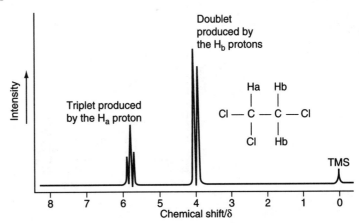

The H_b protons' peak will be split into a doublet because they will experience the applied magnetic field modified by the local field produced by H_a, which may either align with or oppose the applied field.

The H_a proton peak will be split into a triplet because it will experience the applied magnetic field, modified by **both** H_b protons. Depending on whether their spins are up or down, the combinations will be:
1. Both aligned with the field (**A**).
2. One aligned with the field and one against (two combinations) (**B**).
3. Both aligned against the field (**C**).

Figure 74

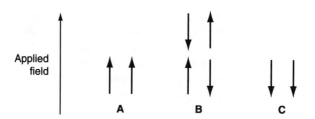

The single H_a proton will therefore interact with a magnetic field modified in three ways, with the second way (B) twice as likely as the others. Thus the H_a peak becomes a $1:2:1$ triplet.

Rules concerning spin-spin splitting have been formulated and a useful pattern emerges. In general, if there are n protons, in a different environment adjacent to the resonating group (or group of protons), the absorption will be split in a multiplet of $(n + 1)$ lines. This is shown below in Fig. 75 in a scheme called Pascal's triangle. This will help you interpret high-resolution NMR spectra of complex molecules.

Figure 75
Pascal's triangle.

Number of chemically equivalent protons causing splitting	Relative intensities of lines in the splitting pattern
0	1
1	1 1
2	1 2 1
3	1 3 3 1
4	1 4 6 4 1
5	1 5 10 10 5 1
6	1 6 15 20 15 6 1

You will find Fig. 75 useful in interpreting an NMR spectrum in the next exercise. Figure 76 shows the high-resolution spectrum for ethanol. Compare this with the low-resolution spectrum given in Exercise 114, page 127 and then attempt the exercise that follows.

Figure 76
High-resolution NMR
spectrum of 100% pure
ethanol.

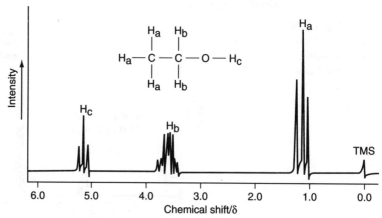

EXERCISE 130
Answers on page 216

The structure of ethanol given in Fig. 76 identifies three different types of proton – H_a, H_b and H_c.

Study it together with the high-resolution spectrum for ethanol, also in Fig. 76, and decide, with reasons, which of the protons is responsible for each of the peaks and splitting patterns shown.

We now show how the addition of $^2_1H_2O$ (D_2O) may be used to identify labile protons, i.e. those liable to displacement or change particularly in compounds containing —OH groups.

EXERCISE 131

Answers on page 216

The NMR spectrum for ethanol contaminated with water is shown below. The structural formula for identifying the different types of protons is also given. Study both and attempt the questions that follow.

Figure 77
High-resolution NMR spectrum of ethanol contaminated with water.

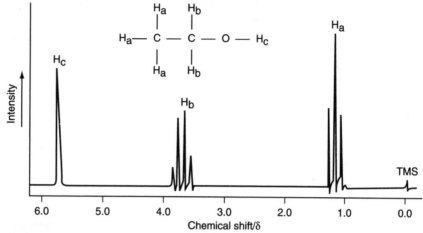

a What has happened to the pattern produced by:
  i) The H_c proton when water is present,
  ii) The H_b proton when water is present?
b If traces of water are present in ethanol, rapid proton exchange occurs:

$$C_2H_5OH^* + HOH = C_2H_5OH + HOH^*$$

  i) How does this proton exchange shown in the equation above explain your answer to (a)?
  ii) When water carrying the isotope of hydrogen $^2_1H_2O$ (D$_2$O) is added to the sample, the peak at 5.7 δ disappears completely. How does this confirm your answer to (b)(i)?

The following teacher-marked exercise gives A-level examination questions you should attempt.

EXERCISES

Teacher-marked

1. **a** In the NMR spectra of organic compounds, protons resonate at different chemical shift values (δ). Explain why the chemical shift of protons in the methyl group of methylbenzene (2.3 δ), is significantly lower than that of the hydroxy proton in phenol (7.0 δ).
   b The NMR spectrum shown below was obtained from the compound G of formula $C_4H_{11}N$.

Figure 78

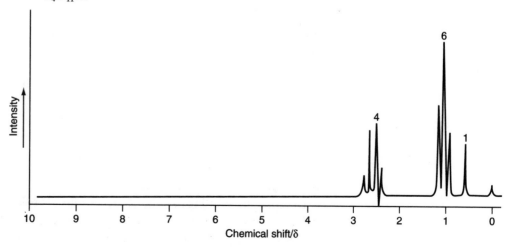

By using Table 19 (page 126) of chemical shift values as necessary, try to identify the compound, explaining how you arrive at your conclusion.

c Explain why, in medicine, NMR spectroscopy is particularly useful in such instruments as body scanners, compared with other spectroscopic techniques.

2. The following NMR spectra were produced by three isomeric compounds of formula $C_4H_{10}O$.

Figure 79
Spectrum 1.

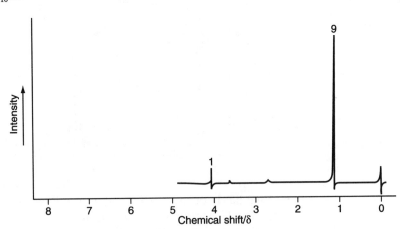

Figure 80
Spectrum 2.

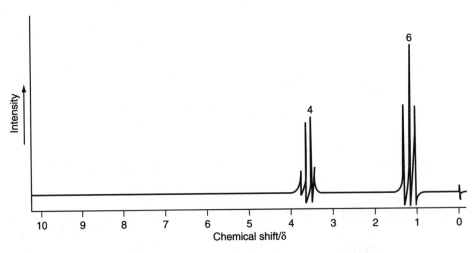

Figure 81
Spectrum 3.

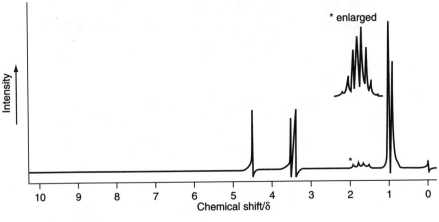

By studying each spectrum in turn:

a Describe the splitting pattern of each group of peaks.

b Identify the arrangement of hydrogen atoms responsible for each group.

c Draw the displayed (full structural) formula for each of the compounds.

X-RAY CRYSTALLOGRAPHY

In ILPAC 3, Bonding and Structure, we mentioned X-ray diffraction as a powerful technique for determining the arrangement of atoms and ions in crystals of simple substances. In this book, you learned that the technique reveals crystalline character in some polymers and has been used to work out very complex structures such as those of DNA and a number of proteins. So far you have focussed on the results of X-ray crystallography; now you consider the technique itself and the theory behind it. The derivation and use of the Bragg equation may not be a requirement of your particular syllabus. Check which of the objectives below apply to you and skip the rest.

OBJECTIVES When you have finished this appendix you should be able to:
- state the general conditions in which **diffraction** occurs giving rise to **diffraction patterns**, and give some specific examples;
- describe, in non-mathematical terms, how **X-ray diffraction patterns** arise;
- derive and use the **Bragg equation**;
- outline one method of producing X-ray diffraction patterns;
- describe briefly how X-ray diffraction helps to determine structures.

■ A5.1 Diffraction patterns

All types of wave motion, including X-rays, interact with regularly spaced arrays of particles, provided that the **wavelength is about the same size as the spacing in the arrays**. (The wavelength can be up to about 1000 times smaller than the spacing distance, but cannot be larger.) This interaction is known as **diffraction**, and the result of the interaction can be observed in the form of a **diffraction pattern**.

If you have studied physics you have probably seen some examples of diffraction patterns. If not, your teacher may be able to suggest some suitable book references.

The next exercise is about diffraction of different types of wave motion.

EXERCISE 132
Answers on page 217

For each wave motion in the list below, find out the approximate wavelength and choose the array of particles most likely to give a diffraction pattern.

Wave motion	**Array**
Microwaves	Crystal
Water waves on a pond	Crystal model made from 5 cm spheres
Light (visible)	Row of fence posts or railings
X-rays	Piece of closely woven cloth

Some examples of X-ray diffraction patterns, recorded on photographic film, are shown in Fig. 82. Each pattern is formed by the interaction of a beam of X-rays with a crystal or crystals, but the method used is slightly different in each case.

Figure 82
Some examples of X-ray diffraction patterns.

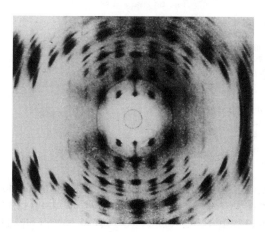
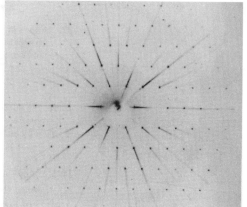

You may also see X-ray diffraction patterns consisting of a series of concentric rings. Later in this appendix, you will read more about the different methods of obtaining X-ray diffraction patterns; first we consider how X-rays interact with crystals.

■ A5.2 X-rays as wave motion

X-rays are a form of wave motion. In any wave motion, a disturbance of some sort travels in the direction of the wave in such a way that the size of the disturbance varies like a sine curve. Figure 83 shows how the disturbance at any given **moment** varies with distance from the source of the wave. Figure 84 shows how the disturbance at any given **point** in the path of the wave varies with time.

Figure 83

Figure 84
(right)

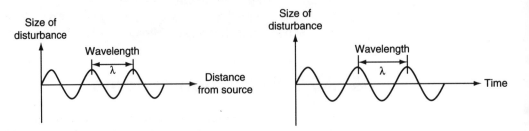

When ripples travel across a pond, the disturbance is a **vertical** movement of water which causes a floating object to bob up and down as the waves pass it. In X-rays the disturbance is an electric field at right angles to the direction of the wave, as it is in all forms of electromagnetic waves including light, infra-red, ultraviolet, radio, radar, etc. These forms of electromagnetic waves differ only in their wavelengths. X-rays have very short wavelengths; radio-waves have long wavelengths.

■ A5.3 Reinforcement and cancellation of waves

The principle behind the formation of all diffraction patterns is that each particle in the grid or array scatters some of the radiation giving a whole series of new waves travelling in all directions. Most of these new waves cancel each other out but, in certain directions, they reinforce each other, as shown in Fig. 85. The reinforced waves give rise to the spots on diffraction patterns.

Figure 85

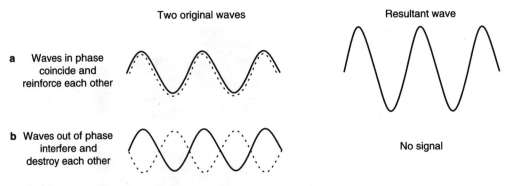

In the next section you learn how measurement of the angles between incident wave and reinforced diffracted waves can give information about crystal structure. If the Bragg equation is not a requirement of your syllabus, skip it and proceed to Section A5.5.

■ A5.4 The Bragg equation

W H and W L Bragg, father and son, simplified the mathematical treatment of the diffraction of X-rays, by considering it as **reflection** from **equally spaced planes of atoms** in the crystal. Some examples of such planes are shown in two dimensions in Fig. 86 for a simple cubic lattice. Figure 87 shows, in three dimensions, the more important planes in the three types of cubic lattice.

Figure 86

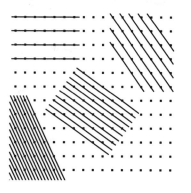

Figure 87
Planes of atoms in cubic lattices.

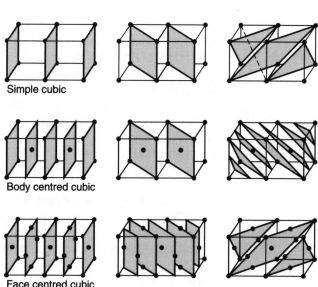

Simple cubic

Body centred cubic

Face centred cubic

The Braggs worked out an equation, known as the Bragg equation or Braggs' Law, which enabled them to calculate the spacings between the reflecting planes from measurements of the angles between reflected (or diffracted) beams and the undeflected beam.

You should study the derivation of the Bragg equation in your textbooks so that you can do Exercise 133. However, as you read, you should bear in mind some points that are not always made clear:

1. You should regard the incident beam of X-rays as a **single** wave motion which is split (or reflected) by successive layers of atoms into a **number** of reflected beams that may or may not reinforce each other.
2. Most of the incident beam passes straight through the crystal undeflected. The reflected beams forming the diffraction pattern are relatively weak.
3. n in the Bragg equation is usually assumed to have a value of 1, but reflected beams also arise for $n = 2, 3$, etc., becoming weaker as n becomes larger.

EXERCISE 133

Answers on page 217

Figure 88 shows the diffraction of X-rays by layers of atoms in a crystal. The incident beam which is, of course, very wide compared with the distance between layers of atoms, can be represented by any straight line in the direction of the beam. Two such lines, known as rays, are shown. Each layer of atoms gives a weak reflected wave, each represented by a single ray. In the example shown, the reflected waves reinforce each other, giving a reflected beam of X-rays which can be detected.

Figure 88

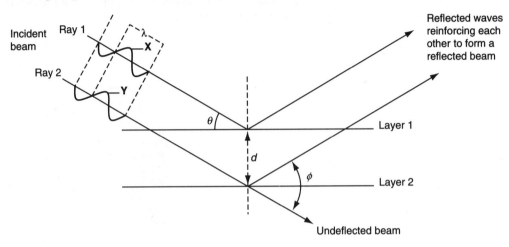

a Although the path length of the two rays is different, the diffracted rays are still in phase. On the diffracted rays draw the waves showing how the relative positions of the peaks X and Y have changed.

b During the X-ray diffraction of a sodium chloride crystal, a value of θ is obtained when the diffracted rays are still in phase, that is, they reinforce each other. What condition must be fulfilled for this to happen? Show any geometrical construction lines on the diagram.

Not part of A-level question

c Since the reflecting layers may not be parallel to the crystal surface, it is not possible to measure θ directly. The angle ϕ, however, is easily measured. How is ϕ related to θ?

d Sir Lawrence Bragg, who carried out the first X-ray investigation of sodium chloride, found that there were several values of θ, each giving rise to a peak on the detector trace, for which there was reinforcement of the diffracted rays. Also, he found that three different detector traces could be obtained according to the position of the crystal.

Why are there several peaks on each trace in Fig. 89?

Figure 89

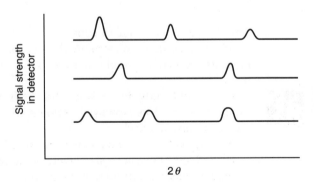

e Why is it possible to obtain three different traces?

Now that you have worked through the derivation of the Bragg equation, you should be able to do some calculations based on it.

EXERCISE 134
Answer on page 217

What is the distance between successive planes which reflect fairly strongly X-rays of wavelength 0.0583 nm at an angle of 9° to the planes?

EXERCISE 135
Answer on page 217

The angle between the undeflected beam and a reflected beam of X-rays (wavelength 0.0635 nm) was found to be 47° 30′. Calculate three possible spacings between the reflecting planes. How, in practice, might you be able to tell which of the three is the one you want?

EXERCISE 136
Answer on page 218

The Bragg equation can be used to determine the wavelength of X-rays, using a crystal of known dimensions. If the incident angle is 12° to a set of reflecting planes 0.198 nm apart, what is the wavelength of the X-rays? (Assume $n = 1$)

■ A5.5 Methods for obtaining X-ray diffraction patterns

Find out from your textbook(s) how X-ray diffraction instruments are used. For examination purposes you need only know an outline of **one** method but, as most books describe several, we summarise the points of difference so that you can choose sensibly.
1. Nowadays the X-ray beam is always monochromatic (i.e. single wavelength). Early methods used a range of wavelengths which gave very complex patterns, difficult to interpret.
2. The X-ray beam may be directed at a single crystal or a powder consisting of many micro-crystals. A powder gives simple patterns consisting of rings rather than spots.
3. The diffracted beams may be detected by photographic film (flat or curved round the crystal) or by a Geiger–Müller tube which mechanically scans the space round the crystal. The G–M tube has the advantage of measuring the intensities of the beams as well as the angles.

If it is available, watch the section of the ILPAC video programme 'Instrumental Techniques' that shows one particular instrument in use for X-ray crystallography. (Also shown is an unrelated technique called nuclear magnetic resonance because it complements X-ray diffraction. NMR, as you already know from the earlier section on NMR spectroscopy, is particularly useful in determining the positions of hydrogen atoms, which are not shown clearly by X-ray diffraction.) The Royal Society of Chemistry video 'Modern chemical techniques' also includes a clip on X-ray crystallography.

EXERCISE 137
Answers on page 218

a In X-ray diffraction, why is the crystal rotated?
b Why do the spots in an X-ray diffraction pattern vary in intensity?
c Why do hydrogen atoms make very little contribution to X-ray diffraction patterns?

Now that you have looked at some diffraction patterns and the methods by which they can be obtained, we consider briefly how they are interpreted.

■ A5.6 Detailed analysis of X-ray diffraction patterns

Only general indications of the type of lattice can be obtained by simple inspection of diffraction patterns. However, it is possible, by use of a computerised mathematical technique called Fourier synthesis, to obtain detailed information about the electron distribution in the reflecting planes.

Another form of presentation, which is even more useful, is an electron contour map or electron-density map. You have interpreted such maps in ILPAC 3, Bonding and Structure, to identify ionic bonding and to obtain values of ionic radius. In the final exercise you interpret another electron-density map.

EXERCISE 138

Answers on page 218

Figure 90

The electron-density map below refers to a substance with the molecular formula $C_8H_8O_3$.

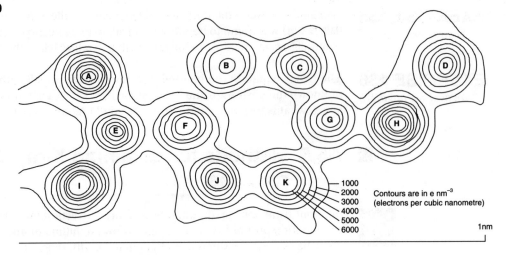

1000
2000
3000
4000
5000
6000

Contours are in e nm^{-3}
(electrons per cubic nanometre)

1nm

a What type of bonding is present? Explain.
b Identify the atoms lettered **A–K**.
c What information does the map give about the positions of hydrogen atoms?

ANSWERS

(Answers to questions from examination papers are provided by ILPAC and not by the examination boards.)

EXERCISE 1 **a**

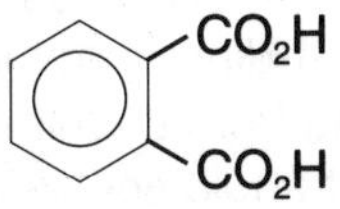

    b $CH_3CH_2CH_2CO_2H$
    c $CH_3CH(OH)CO_2H$
    d

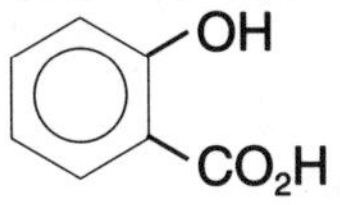

    e $CH_3CH_2CH(CH_3)CH_2CO_2H$
    f $(CO_2H)_2$

EXERCISE 2 **a** Propanoic acid.
    b Phenylethanoic acid.
    c 3-Methylbutanoic acid.
    d Benzene-1,4-dicarboxylic acid.
    e Pentanedioic acid.

EXERCISE 3 **a** Hydrogen evolved as the sodium displaces the hydrogen of the —OH group.
    b No reaction.
    c Hydrogen chloride evolved.
    d A yellow precipitate.
    e A pale yellow precipitate if the CH_3CO— or CH_3CHOH group is present.
    f Either no colour as shown by alcohols and carbonyl compounds or a violet colour shown by the —OH group in phenols.
    g Either a neutral green solution shown by alcohols and carbonyl compounds or an acidic red solution shown by the —OH group in phenols.

EXPERIMENT 1
Specimen results
Results Table 1
Reactions of ethanoic acid

Reaction		Observations
A.	pH of aqueous solution	Orange–red – pH 3–4
B.	Reaction with sodium hydrogen-carbonate solution	Gas evolved which turns limewater milky – CO_2
C.	Reaction with sodium	Gas evolved which popped with a lighted splint – H_2
D.	Reaction with phosphorus pentachloride	Steamy gas evolved which gave white fumes with ammonia – HCl
E.	Reaction with 2,4-dinitro-phenylhydrazine	No change
F.	Triiodomethane reaction	No change
G.	Action of iron(III) chloride	Red colour at first – brown ppt. on boiling

Questions
1. Only the reactions with sodium and phosphorus pentachloride corresponded with the predictions.
2. Ethanoic acid resembles hydroxy compounds more closely than carbonyl compounds. It reacts with sodium and phosphorus pentachloride, typical reactions of alcohols, but does not react with 2,4-dinitrophenylhydrazine.
3. Ethanoic acid releases carbon dioxide from sodium hydrogencarbonate, but phenol does not. Also, the pH of aqueous ethanoic acid is lower than that of aqueous phenol of similar concentration.
4. The presence of the large, non-polar benzene ring outweighs the influence of the polar carboxyl group, which participates in hydrogen-bonding with water.

EXERCISE 4

a $CH_3CO_2H + NaOH \rightarrow CH_3CO_2^-Na^+ + H_2O$
  sodium ethanoate
(Treat the acid with aqueous alkali)

b $2CH_3CH_2CO_2H + Na_2CO_3 \rightarrow 2CH_3CH_2CO_2^-Na^+$
  sodium propanoate $+H_2O + CO_2$
(Treat acid with aqueous carbonate)

c $CH_3CO_2H + (NH_4)_2CO_3 \rightarrow CH_3CO_2^-NH_4^+$
  ammonium ethanoate $+CO_2 + H_2O$
$CH_3CO_2^-NH_4^+ \rightarrow CH_3CONH_2 + H_2O$
  ethanamide
(Heat solid ammonium carbonate with excess acid – to prevent dissociation of the salt. First stage can be in aqueous solution)

d $C_6H_5CO_2H + PCl_5 \rightarrow C_6H_5COCl + POCl_3 + HCl$
  benzoyl chloride
(Distil solid mixture. No heat required for liquid acids)

A similar reaction occurs with sulphur dichloride oxide (thionyl chloride):

$$C_6H_5CO_2H + SOCl_2 \rightarrow C_6H_5COCl + SO_2 + HCl$$

e $CH_3CH_2CO_2H + 2H_2 \rightarrow CH_3CH_2CH_2OH + H_2O$
(Distil with $LiAlH_4$ in dry ethoxyethane)

f $CH_3CH_2CO_2H + CH_3OH \rightarrow CH_3CH_2CO_2CH_3 + H_2O$
  methyl propanoate
(Heat with a little concentrated sulphuric acid)

EXERCISE 5
Table 1
A comparative summary

Compound	Observation when tested with various reagents			
	Neutral iron(III) chloride	**Sodium hydrogen-carbonate**	**2,4-dinitro-phenyl-hydrazine**	**Tollens reagent**
Ethanoic acid CH_3CO_2H	Red colour	CO_2 evolved	No change	No change
Benzoic acid $C_6H_5CO_2H$	Buff ppt.	CO_2 evolved	No change	No change
Ethanol C_2H_5OH	No change	No change	No change	No change
Phenol C_6H_5OH	Violet colour	No change	No change	No change
Ethanal CH_3CHO	No change	No change	Orange ppt.	Silver mirror

EXERCISE 6 **a** Any one of the following reactions would serve.

To each in turn add neutral iron(III) chloride. With benzoic acid, $C_6H_5CO_2H$, there is a buff precipitate but with benzaldehyde, C_6H_5CHO, there is no change.

To each in turn add sodium hydrogencarbonate. Benzoic acid gives off carbon dioxide but pure benzaldehyde does not. (However, benzaldehyde is often contaminated with its oxidation product, benzoic acid.)

To each in turn add 2,4-dinitrophenylhydrazine solution. Benzaldehyde gives a yellow precipitate but benzoic acid does not.

To each in turn add a solution of silver oxide in aqueous ammonia and warm. Benzaldehyde forms a silver mirror but benzoic acid does not.

b To each in turn add neutral iron(III) chloride solution. Benzoic acid, $C_6H_5CO_2H$, forms a buff precipitate and phenol, C_6H_5OH, forms a violet colour.

Alternatively, to each in turn add sodium hydrogencarbonate. Benzoic acid gives off carbon dioxide but phenol does not.

c To each in turn add neutral iron(III) chloride solution. Ethanoic acid, CH_3CO_2H, forms a red colour and benzoic acid, $C_6H_5CO_2H$, forms a buff precipitate.

EXERCISE 7 **a** i) Ethanoic acid and benzoic acid.
ii) $CH_3CO_2H + NaHCO_3 \rightarrow CH_3CO_2Na + CO_2 + H_2O$
    sodium ethanoate
or $C_6H_5CO_2H + NaHCO_3 \rightarrow C_6H_5CO_2Na + CO_2 + H_2O$
    sodium benzoate

b i) Phenol.
ii) $C_6H_5OH + NaOH \rightarrow C_6H_5ONa + H_2O$
    sodium phenoxide

c i) Ethanol. (You may have seen the following equation for a reaction with alkali but the equilibrium lies very heavily to the left:
$C_2H_5OH + OH^- \rightleftharpoons C_2H_5O^- + H_2O$)
    ethoxide ion
ii) $2C_2H_5OH + 2Na \rightarrow 2C_2H_5ONa + H_2$
    sodium ethoxide

EXERCISE 8 **a** The conventional way of representing the carboxylate ion must be wrong because it does not represent a symmetrical molecule. The $C=O$ bond is shorter than the $C-O$ bond.

b

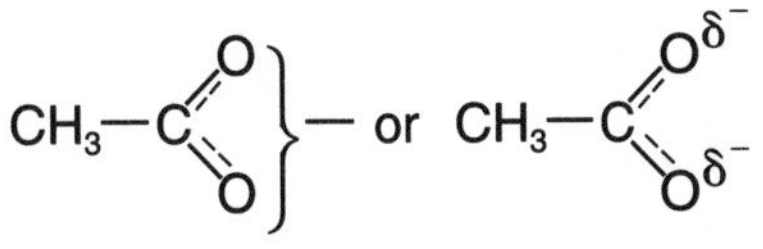

$$CH_3-C{\begin{matrix} O \\ \\ O \end{matrix}}\Big\} \quad \text{or} \quad CH_3-C{\begin{matrix} O^{\delta-} \\ \\ O^{\delta-} \end{matrix}}$$

c p orbitals overlap to form delocalised π orbitals.

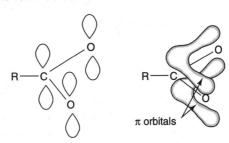

π orbitals

d In ethanoic acid, unlike ethanol, there is an acyl group, RCO, which draws electrons away from the hydroxyl oxygen atom.

$$CH_3-C\diagdown\diagup{}^{O}_{O-H}$$

This electron withdrawal polarises and weakens the O—H bond, which facilitates the loss of a proton.

Also, the ethanoate ion $CH_3CO_2^-$, unlike the ethoxide ion, $C_2H_5O^-$, is greatly stabilised with respect to the un-ionised molecule by delocalisation, with its charge spread between two oxygen atoms. This means that ethanoate ions recombine with protons less readily, which makes ethanoic acid more acidic than ethanol.

EXERCISE 9 **a** The trend in acid strength is:

$$CCl_3CO_2H > CHCl_2CO_2H > CH_2ClCO_2H > CH_3CO_2H$$

Acid strength is increased by the presence of electron-attracting substituents (in this case, chlorine) in the CH_3 group. This is because the presence of the electron-withdrawing substituent further polarises and weakens the O—H bond and also stabilises the carboxylate ion by dispersing some of the negative charge on the oxygen atoms of the ion.

$$\leftarrow C \diagup\diagdown {}^O_O \Big\} -$$

As we might expect, increasing the number of such substituents increases the acidity.

b Decreasing order of acid strength:

$$FCH_2CO_2H > ClCH_2CO_2H > BrCH_2CO_2H > ICH_2CO_2H > CH_3CO_2H$$

The electron-withdrawing power of the halogen atoms is indicated by values of electronegativity where F > Cl > Br > I.

c Trend in acid strength:

$$CH_3CH_2CHClCO_2H > CH_3CHClCH_2CO_2H > CH_2ClCH_2CH_2CO_2H$$

As the distance between the electron-withdrawing substituent (in this case, chlorine) and the carboxyl group increases, the electron-withdrawing effect on the carboxyl group is reduced. This reduces the acidity of the compounds.

d Trend in acid strength: $HCO_2H > CH_3CO_2H > CH_3CH_2CO_2H$

Alkyl groups are electron releasing:

$$H-\overset{\displaystyle H}{\underset{\displaystyle H}{C}}\rightarrow C\diagup\diagdown{}^{O}_{O-H}$$

This has the effect of reducing the polarisation of the O—H bond. The bond is strengthened which makes proton loss more difficult.

Acid strength therefore decreases as the electron-releasing effect increases:

$$H > CH_3 > C_2H_5$$

e Both the benzene ring and the methyl group are electron releasing, but the more important factor in this case is ion stabilisation. The benzoate ion is stabilised to a greater extent than the ethanoate ion because the charge is dispersed by interaction with the delocalised ring system. Consequently, the benzoate ion accepts protons less readily and benzoic acid is the stronger acid.

EXERCISE 10   Table 3 General methods of preparing carboxylic acids

A. Oxidation

1. Primary alcohols

$CH_3CH_2OH + H_2O \rightarrow CH_3CO_2H + 4e^- + 4H^+$

ethanol   ethanoic acid

Reflux with acidified dichromate(VI), $Cr_2O_7^{2-}$

2. Aldehydes

$CH_3CHO + H_2O \rightarrow CH_3CO_2H + 2e^- + 2H^+$

ethanal   ethanoic acid

Reflux with acidified dichromate(VI), $Cr_2O_7^{2-}$

3. Alkyl benzenes

$C_6H_5CH_3 + 2H_2O \rightarrow C_6H_5CO_2H + 6e^- + 6H^+$

methylbenzene   benzoic acid

Reflux with acidified manganate(VII), MnO_4^-

B. Hydrolysis

1. Nitriles

$CH_3CH_2CN + 2H_2O + H^+ \rightarrow CH_3CH_2CO_2H + NH_4^+$

propanenitrile   propanoic acid

Reflux with dilute acid (or alkali to give a salt)

2. Amides

$CH_3CONH_2 + H_2O + H^+ \rightarrow CH_3CO_2H + NH_4^+$

ethanamide   ethanoic acid

Reflux with dilute acid (or alkali to give a salt)

3. Acyl halides

$CH_3COCl + H_2O \rightarrow CH_3CO_2H + HCl$

ethanoyl chloride   ethanoic acid

Vigorous reaction in cold water

4. Anhydrides

$(CH_3CO)_2O + H_2O \rightarrow 2CH_3CO_2H$

ethanoic anhydride   ethanoic acid

Heat with water (or dilute alkali)

5. Esters

$CH_3CH_2CO_2CH_3 + H_2O \rightarrow CH_3CH_2CO_2H + CH_3OH$

methyl propanoate   propanoic acid   methanol

Heat with dilute acid (or alkali)

Preparation of hydroxycarboxylic acids

Cyanohydrin reaction

$CH_3CHO + HCN \rightarrow CH_3CH(OH)CN$

ethanal   2-hydroxypropane nitrile

$CH_3CH(OH)CN + 2H_2O + H^+ \rightarrow CH_3CH(OH)CO_2H + NH_4^+$

2-hydroxypropanoic acid

} 2 stages

Mix with NaCN and dilute acid, then reflux

$CH_3COCH_3 + HCN \rightarrow (CH_3)_2C(OH)CN$

propanone   2-hydroxy-2-methylpropane nitrile

$(CH_3)_2C(OH)CN + 2H_2O + H^+ \rightarrow (CH_3)_2C(OH)CO_2H + NH_4^+$

2-hydroxy-2-methylpropanoic acid

} 2 stages

Mix with NaCN and dilute acid, then reflux

Industrial methods

A. Methanoic acid

$NaOH(aq) + CO(g) \rightarrow HCO_2^-Na^+(aq)$

$HCO_2^-Na^+(aq) + H^+(aq) \rightarrow HCO_2H + Na^+(aq)$

200°C, high pressure

Acidify

B. Ethanoic acid

$alkane + O_2 \rightarrow CH_3CO_2H$

Mixture with air at high temperature and pressure passed over Co^{2+} catalyst

C. Benzoic acid

$C_6H_5CH_3 + 1\frac{1}{2}O_2 \rightarrow C_6H_5CO_2H + H_2O$

methylbenzene   benzoic acid

Mixture with air at high temperature and pressure passed over tin(IV) vanadate(V) catalyst

D. Ethanedioic acid

$2HCO_2Na \rightarrow (CO_2Na)_2 + H_2$

sodium methanoate   sodium ethanedioate

$(CO_2Na)_2 + 2H^+(aq) \rightarrow (CO_2H)_2 + 2Na^+$

ethanedioic acid

} 2 stages

Heat solid

Treat with dilute acid

EXERCISE 11

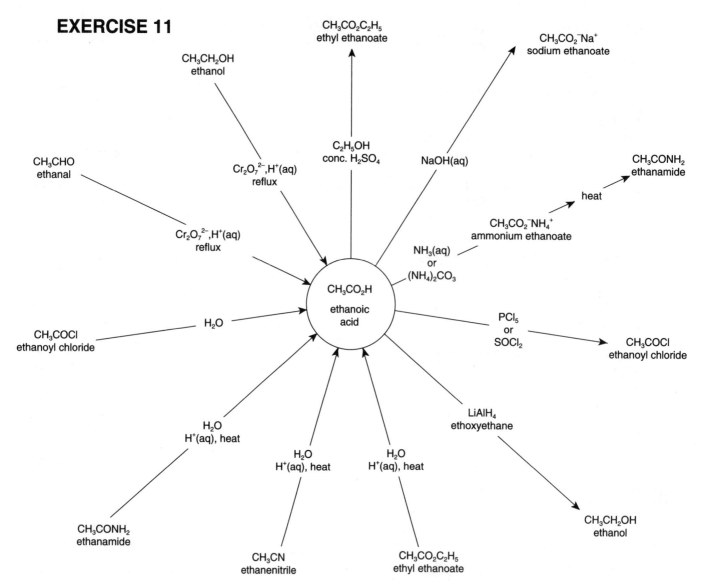

EXERCISE 12

Step 1: conc. HNO_3/H_2SO_4, 60 °C, substitution.
Step 2: Cl_2 (inert solvent), ultraviolet light, substitution.
Step 3: OH^-(aq), heat, substitution.
Step 4: $Cr_2O_7^{2-}$, H^+(aq), reflux, oxidation.
Step 5: C_2H_5OH, H_2SO_4, heat, esterification.
Step 6: Sn/HCl(aq), heat, reduction.

EXERCISE 13

a $C_6H_5NH_2 \xrightarrow[<10\,°C]{NaNO_2/H^+(aq)} C_6H_5N_2^+ \xrightarrow[Heat]{KCN(aq)/CuCN} C_6H_5CN \xrightarrow[Reflux]{H^+(aq)} C_6H_5CO_2H$

b $CH_3CHO \xrightarrow{KCN/H^+(aq)} CH_3CH(OH)CN \xrightarrow[Reflux,\ acidify]{OH^-(aq)} CH_3CH(OH)CO_2H$

c $CH_3(CH_2)_3I \xrightarrow[Reflux]{KCN/C_2H_5OH} CH_3(CH_2)_3CN \xrightarrow[Reflux,\ acidify]{OH^-(aq)} CH_3(CH_2)_3CO_2H$

d $CH_3(CH_2)_3CO_2H \xrightarrow[ethoxyethane]{LiAlH_4} CH_3(CH_2)_4OH \xrightarrow[Reflux]{P/I_2} CH_3(CH_2)_4I$

EXERCISE 14

$$CH_3CO_2H \ + \ \langle \bigcirc \rangle{-}CH_2OH \ \xrightarrow[\text{Heat}]{\text{conc. } H_2SO_4} \ CH_3CO_2CH_2{-}\langle \bigcirc \rangle + H_2O$$

ethanoic acid   phenylmethanol   phenylmethylethanoate

EXERCISE 15   a   Hexanedioic acid is an intermediate in the manufacture of nylon.
b   Ethanoic acid is used in aqueous solution in vinegar.
c   Benzoic acid (E210) and some of its salts (e.g. E211 sodium benzoate) are used as preservatives in foodstuffs.
d   Benzene-1,4-dicarboxylic acid is used in the manufacture of Terylene (Crimplene, Dacron, polyester, Trevira).

EXERCISE 16   a   i)   Ethyl methanoate.
ii)   Propanoyl chloride.
iii)   Propanamide.
iv)   Ethanoic propanoic anhydride.
v)   Ethyl benzoate.
b   i) $CH_3CO_2CH_2CH_2CH_3$ (or $CH_3CH_2CH_2OCOCH_3$)
ii) $CH_3CH_2CH_2CONH_2$

iii) $C_6H_5COCCH_3$ (or $C_6H_5CO_2COCH_3$)
           ‖ ‖
           O O

iv) $C_6H_5CO_2C_6H_5$ (or $C_6H_5OCOC_6H_5$)

EXPERIMENT 2

Specimen results

Results Table 2

Reaction	Observations
A. **Reaction with water** Product tested with ammonia Product tested with iron(III) chloride	Moderate reaction giving steamy fumes White fumes, suggesting HCl is produced Red coloration, suggesting ethanoic acid is formed
B. **Reaction with ethanol** Product tested with ammonia Smell of product	Vigorous reaction with crackling and spitting, giving steamy fumes White fumes, suggesting HCl is produced Sweetish smell, suggesting an ester is produced
C. **Reaction with ammonia**	Violent reaction, with crackling and spitting, giving a lot of white fumes (NH_4Cl)
D. **Reaction with phenylamine**	Violent reaction with crackling and spitting, giving a lot of white fumes. Some white solid around lower walls of beaker

Questions   1. $CH_3COCl + H_2O \rightarrow CH_3CO_2H + HCl$
                           ethanoic acid
$CH_3COCl + C_2H_5OH \rightarrow CH_3CO_2C_2H_5 + HCl$
                           ethyl ethanoate
$CH_3COCl + 2NH_3 \rightarrow CH_3CONH_2 + NH_4Cl$
                           ethanamide
$CH_3COCl + C_6H_5NH_2 \rightarrow CH_3CONHC_6H_5 + HCl$ (HCl reacts with excess amine)
                           N-phenylethanamide

2.   The order of reactivity of the nucleophiles corresponds with the order of basic strength.

$$NH_3 > C_6H_5NH_2 > C_2H_5OH > H_2O$$

The more basic the nucleophile, the more readily available is the lone pair of electrons to make a bond with ethanoyl chloride, as shown in Fig. 4 on page 26.

EXERCISE 17

Table 4

Reactions of carboxylic acid derivatives with nucleophiles

A. **With water**	**Conditions**
1. Acyl halides $C_2H_5COCl + H_2O \rightarrow C_2H_5CO_2H + HCl$ propanoyl chloride propanoic acid	Reacts cold
2. Acyl anhydrides $(CH_3CO)_2O + H_2O \rightarrow 2CH_3CO_2H$ ethanoic ethanoic anhydride acid	Warm
3. Esters $C_2H_5CO_2CH_3 + H_2O \rightarrow C_2H_5CO_2H + CH_3OH$ methyl propanoic methanol propanoate acid	Reflux with dilute acid, (or alkali gives $C_2H_5CO_2^-$)
4. Amides $CH_3CONH_2 + H_2O \rightarrow CH_3CO_2H + NH_3$ ethanamide ethanoic acid	Reflux with dilute acid ($\rightarrow NH_4^+$) or alkali ($\rightarrow CH_3CO_2^-$)
B. **With alcohols and phenols**	
1. Acyl halides $C_2H_5COCl + C_2H_5OH \rightarrow C_2H_5CO_2C_2H_5 + HCl$ propanoyl chloride ethyl propanoate	Reacts cold
2. Acyl anhydrides $CH_3CO_2COCH_3 + C_6H_5OH \rightarrow CH_3CO_2C_6H_5 + CH_3CO_2H$ ethanoic phenyl ethanoic anhydride ethanoate acid	Reflux
C. **With ammonia**	
1. Acyl halides $C_2H_5COCl + 2NH_3 \rightarrow C_2H_5CONH_2 + NH_4Cl$ propanoyl chloride propanamide	Reacts cold
2. Acyl anhydrides $C_2H_5CO_2COC_2H_5 + NH_3 \rightarrow C_2H_5CONH_2 + C_2H_5CO_2H$ propanoic anhydride propanamide propanoic acid	Reacts cold
3. Esters $CH_3CO_2C_2H_5 + NH_3 \rightarrow CH_3CONH_2 + C_2H_5OH$ ethyl ethanoate ethanamide ethanol	Reacts cold
D. **With amines**	
1. Acyl halides $C_2H_5COCl + CH_3NH_2 \rightarrow C_2H_5CONHCH_3 + HCl$ propanoyl methyl- *N*-methylpropanamide chloride amine	HCl reacts with excess amine
2. Acyl anhydrides $(CH_3CO)_2O + C_3H_7NH_2 \rightarrow CH_3CONHC_3H_7 + CH_3CO_2H$ ethanoic propyl- *N*-propyl- ethanoic anhydride amine ethanamide acid	Heat
3. Esters $CH_3CO_2C_2H_5 + C_2H_5NH_2 \rightarrow CH_3CONHC_2H_5 + C_2H_5OH$ ethyl ethyl- *N*-ethyl- ethanol ethanoate amine ethanamide	Reflux. Heat

EXERCISE 18   **a**

$$H_2O: \quad \underset{Cl}{\overset{CH_3}{\underset{|}{C}}} \overset{\delta+}{=}\overset{\delta-}{O} \longrightarrow H-\underset{H}{\overset{CH_3}{\underset{|}{\overset{\oplus}{O}}}}-\underset{Cl}{\overset{|}{C}}-\overset{\ominus}{O} \longrightarrow H-O-\overset{CH_3}{\underset{|}{C}}=O + HCl$$

b The first stage is nucleophilic addition; the second is elimination.

c i) The first stage in each type of reaction is the same, viz. nucleophilic addition.

For carbonyl compounds:
$$HZ: \quad \underset{R}{\overset{R}{\underset{|}{C}}}=O \longrightarrow HZ^{\oplus}-\underset{R}{\overset{R}{\underset{|}{C}}}-O^{\ominus}$$

For acyl chlorides:
$$HZ: \quad \underset{Cl}{\overset{R}{\underset{|}{C}}}=O \longrightarrow HZ^{\oplus}-\underset{Cl}{\overset{R}{\underset{|}{C}}}-O^{\ominus}$$

The second stage is different because the chlorine atom is electron withdrawing whereas the alkyl group, R, is electron donating. The chlorine atom, therefore, tends to break away and link with a hydrogen atom to eliminate HCl. The overall process is substitution of Cl by Z.

$$\overset{\oplus}{Z}-\underset{\underset{Cl}{\overset{|}{H}}}{\overset{R}{\underset{|}{C}}}-O^{\ominus} \longrightarrow Z-\overset{R}{\underset{|}{C}}=O + HCl$$

By contrast, the intermediate from a carbonyl compound transfers a proton from the attached nucleophile to disperse the charge. Elimination of water follows by means of another proton transfer.

$$\overset{\oplus}{Z}-\underset{\underset{R}{\overset{|}{H}}}{\overset{R}{\underset{|}{C}}}-O^{\ominus} \longrightarrow Z-\underset{R}{\overset{R}{\underset{|}{C}}}-OH \longrightarrow {}^*Z=\overset{R}{\underset{R}{\overset{|}{C}}} + H_2O$$

*This Z has one less H atom than Z in the intermediate

The overall reaction is usually called condensation, or addition–elimination but, for the purposes of comparison, could be regarded as substitution of O by Z.

ii) Like acyl halides, primary alkyl halides also undergo substitution of Cl by nucleophiles. However, for primary alkyl halides, the intermediate adduct cannot bond fully with the nucleophile because the central carbon atom cannot form five bonds. Instead, the approach of the nucleophile and departure of the halogen are simultaneous.

$$Z:^- + \underset{H}{\overset{H}{\underset{|}{C}}}\overset{\delta+}{}-X^{\delta-} \longrightarrow \left[Z\cdots \underset{H}{\overset{H\ \ H}{C}}\cdots X \right]^- \longrightarrow Z-\underset{H}{\overset{H}{\underset{|}{C}}}\diagdown H + X^-$$

activated complex

EXERCISE 19 Hydrolyse the ester by refluxing with dilute acid in which the water contains the oxygen-18 isotope. Separate the products by fractional distillation. If the radioactive isotope appears in the carboxylic acid, this indicates that fission (1) occurs.

(1)

$$R-CO + O-R' + H_2O^* \longrightarrow RCO_2^*H + R'OH \qquad O^* = {}^{18}O$$

On the other hand, if the isotope appears in the alcohol, this indicates that fission (2) occurs.

(2)

$$R-CO-O + R' + H_2O^* \longrightarrow RCO_2H + R'O^*H$$

In general, it is found that acyl-oxygen fission (1) is predominant.

EXERCISE 20

Table 5

Reduction of carboxylic acid derivatives

1.	Acyl halides	e.g. $CH_3COCl + H_2 \rightarrow CH_3CH_2OH + HCl$
2.	Acyl anhydrides	e.g. $(CH_3CO)_2O + 4H_2 \rightarrow 2CH_3CH_2OH + H_2O$
3.	Esters	e.g. $CH_3CH_2CO_2CH_3 + 2H_2 \rightarrow CH_3CH_2CH_2OH + CH_3OH$
4.	Amides	e.g. $CH_3CONH_2 + 2H_2 \rightarrow CH_3CH_2NH_2 + H_2O$

Lithium tetrahydridoaluminate(III), $LiAlH_4$, in dry ethoxyethane can reduce all four types of derivative (as well as the parent carboxylic acid). (For acyl halides the safer alternative is aqueous sodium tetrahydridoborate(III), $NaBH_4$.)

EXERCISE 21 **a** $CH_3CH_2CONH_2 + Br_2 + 4OH^- \rightarrow CH_3CH_2NH_2 + 2Br^- + CO_3^{2-} + 2H_2O$
propanamide   ethylamine

b This reaction removes one atom from a carbon chain and is the vital step in one method of descending a homologous series.

EXERCISE 22 $CH_3CONH_2 \rightarrow CH_3CN + H_2O$
ethanamide   ethanenitrile
Ethanamide is distilled over phosphorus pentoxide.

EXERCISE 23 **a** Ethanoyl chloride reacts more vigorously than chloroethane with nucleophiles. The reactions between ethanoyl chloride and nucleophiles such as water and ammonia take place directly and violently.
    The reaction between chloroethane and water is very slow and the reaction with ammonia only takes place under more extreme conditions by heating to 100 °C with ammonia dissolved in ethanol in a sealed tube.

b The carbon atom in a $-COCl$ group is attached to two strongly electronegative atoms and, therefore, carries a larger partial positive charge than the carbon atom in a $-CCl$ group, which is attached to only one electronegative atom.

c Since acyl halides carry a larger positive charge on the carbon atom of the $-COCl$ group, they will be more susceptible to nucleophilic attack than alkyl halides. This effect is enhanced by the possibility of a shift of the π electrons in the $C=O$ bond, which also makes it possible for the carbon atom to bond with a nucleophile **before** the Cl atom is released, i.e. the intermediate is formed more readily.

$$\begin{array}{c} \diagdown \\ \diagup \end{array} C{=}O \longrightarrow -\overset{\oplus}{\underset{|}{C}}-O^{\ominus}$$

EXERCISE 24 In an amide, the lone pair of electrons on the nitrogen atom becomes incorporated in a delocalised system, and this reduces its ability to accept a proton. No such delocalisation occurs in an amine which, therefore, accepts a proton more readily, i.e. amines are stronger bases than amides.

EXERCISE 25
a Ethanamide < phenylamine < ammonia < butylamine
b i) In phenylamine, $C_6H_5NH_2$, the lone pair of electrons on the nitrogen becomes incorporated into the benzene ring by delocalisation.

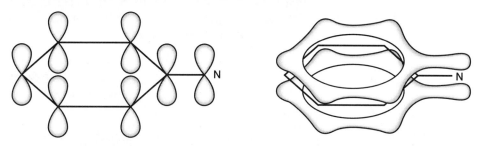

This reduces the availability of the lone pair of electrons to accept a proton, i.e. the base strength is reduced.

In butylamine the $-C_4H_9$ group, like all alkyl groups, is electron releasing.

$$C_4H_9 \rightarrow NH_2$$

This means that the lone pair of electrons on the nitrogen atom is more available to accept a proton and butylamine is therefore a stronger base than phenylamine. Also, the delocalisation in phenylamine means that, unlike butylamine, phenylamine is relatively more stable than its conjugate acid ion. It follows that there will be a reduced tendency to attract a proton and consequently basic properties will be lessened.

ii) In ethanamide, CH_3CONH_2, the lone pair of electrons on the nitrogen atom is incorporated into a delocalised system:

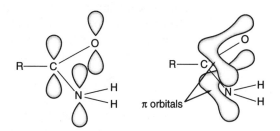

This reduces the availability of the lone pair to accept a proton, i.e. the base strength is reduced.

In butylamine, as explained in i), the electron-donating effect of the C_4H_9 group makes the lone pair of electrons on the nitrogen atom more available to accept a proton, and butylamine is therefore a stronger base than phenylamine.

Also, the delocalisation in ethanamide means that, unlike butylamine, ethanamide is relatively more stable than its conjugate acid ion. This means there will be a reduced tendency to attract a proton and, consequently, basic properties will be lessened.

iii) In both ethanamide and phenylamine, the lone pair of electrons is incorporated into a delocalised system. However, the electron-withdrawing effect of the carbonyl group must be greater than that of the benzene ring, accounting for the weaker base strength of ethanamide compared with phenylamine.

EXERCISE 26   **a** To each in turn add water, dropwise. Ethanoyl chloride reacts violently to release steamy fumes of hydrogen chloride. Chlorobenzene and chlorobutane do not react.

$$C_2H_5COCl + H_2O \rightarrow C_2H_5OH + HCl$$

The other two compounds may be distinguished by warming with aqueous sodium hydroxide, neutralising with dilute nitric acid, and adding aqueous silver nitrate. 1-Chlorobutane gives a white precipitate of silver chloride but chlorobenzene does not because it is resistant to hydrolysis.

$$C_4H_9Cl + OH^- \rightarrow C_4H_9OH + Cl^-$$
$$Ag^+(aq) + Cl^-(aq) \rightarrow AgCl(s)$$

b Heat each in turn with sodium hydroxide solution to produce the corresponding alcohols and carboxylate ions. Then acidify each with dilute hydrochloric acid. The appearance of a solid on cooling identifies benzoic acid produced from the ester containing the benzene ring, i.e. ethyl benzoate, $C_6H_5CO_2C_2H_5$.

$$C_6H_5CO_2C_2H_5 + OH^- \rightarrow C_6H_5CO_2^- + C_2H_5OH$$
ethyl benzoate   benzoate   ethanol
                                             ion
$$C_6H_5CO_2^-(aq) + H^+(aq) \rightarrow C_6H_5CO_2H(s)$$

Ethyl ethanoate is also hydrolysed, but ethanoic acid is soluble in water and no precipitate is formed.

c Heat each in turn with sodium hydroxide to produce the corresponding hydroxy compounds and carboxylate ions. Acidify and cool to allow the benzoic acid to separate out. Filter and test the filtrates in one of two ways.

   i)   Add a solution of iodine in potassium iodide, followed by aqueous sodium hydroxide. The appearance of a yellow precipitate indicates the presence of ethanol, produced from the ester containing the ethanoate group, i.e. ethyl benzoate.

$$C_6H_5CO_2C_2H_5 + OH^- \rightarrow C_6H_5CO_2^- + C_2H_5OH$$
                                                           ethanol
$$C_2H_5OH + 4I_2 + 6OH^- \rightarrow HCO_2^- + 5I^- + 5H_2O + CHI_3$$
                                        methanoate   triiodomethane

   ii) Add neutral iron(III) chloride solution. A violet colour indicates the presence of phenol in the filtrate, which must have come from phenyl benzoate.

$$C_6H_5CO_2C_6H_5 + 2OH^- \rightarrow C_6H_5CO_2^- + C_6H_5O^- + H_2O$$
phenyl benzoate
$$C_6H_5CO_2^- + C_6H_5O^- + 2H^+ \rightarrow C_6H_5CO_2H + C_6H_5OH$$
                                             benzoic   phenol
                                             acid
                                             (filtered off)

EXERCISE 27

Table 6
Formation of derivatives of
carboxylic acids

A. **Acyl halides**	
1. From carboxylic acids $CH_3CO_2H + SOCl_2 \rightarrow CH_3COCl + SO_2 + HCl$ ethanoic ethanoyl acid chloride $CH_3CO_2H + PCl_5 \rightarrow CH_3COCl + POCl_3 + HCl$ $3CH_3CO_2H + PCl_3 \rightarrow 3CH_3COCl + H_3PO_3$	Heat mixture
B. Amides	
1. From ammonium salts $CH_3CH_2CO_2{}^-NH_4{}^+ \rightarrow CH_3CH_2CONH_2 + H_2O$ ammonium propanoate propanamide	Heat, preferably with excess propanoic acid to prevent dissociation of salt
2. From acyl halides $C_3H_7COCl + 2NH_3 \rightarrow C_3H_7CONH_2 + NH_4Cl$ butanoyl chloride butanamide	Conc. NH_3(aq) at room temperature
3. From esters $CH_3CO_2C_2H_5 + NH_3 \rightarrow CH_3CONH_2 + C_2H_5OH$ ethyl ethanoate ethanamide	Conc. NH_3(aq) at room temperature
4. From anhydrides $(CH_3CO)_2O + 2NH_3 \rightarrow CH_3CONH_2 + CH_3CO_2{}^-NH_4{}^+$ ethanoic anhydride ethanamide	Conc. NH_3(aq). Reaction 1 follows if heated
C. Esters	
1. From carboxylic acids and hydroxy compounds $CH_3CO_2H + C_2H_5OH \rightarrow CH_3CO_2C_2H_5 + H_2O$ ethanoic acid ethyl ethanoate	Heat with conc. H_2SO_4
2. From acyl halides and hydroxy compounds $C_2H_5COCl + C_6H_5OH \rightarrow C_2H_5CO_2C_6H_5 + HCl$ propanoyl chloride phenylpropanoate	Mix at room temperature
3. From anhydrides and hydroxy compounds $(CH_3CO)_2O + C_2H_5OH \rightarrow CH_3CO_2C_2H_5 + CH_3CO_2H$ ethanoic anhydride ethyl ethanoate	Warm with conc. H_2SO_4

b Sulphur dichloride oxide, $SOCl_2$, is particularly useful in the preparation of acyl halides. The by-products of the reaction are both gases, so that separation and purification of the acyl halide is easier.

EXERCISE 28

For other reactions see
Figure 2

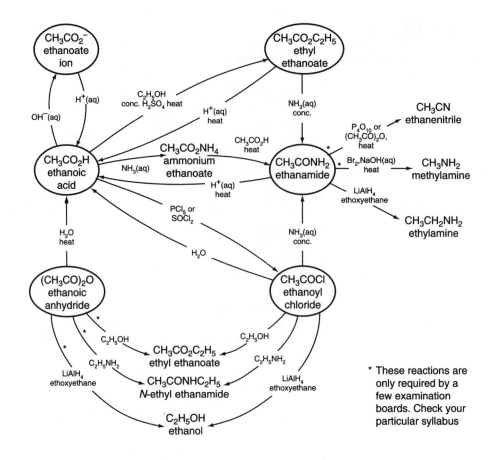

* These reactions are
only required by a
few examination
boards. Check your
particular syllabus

EXERCISE 29

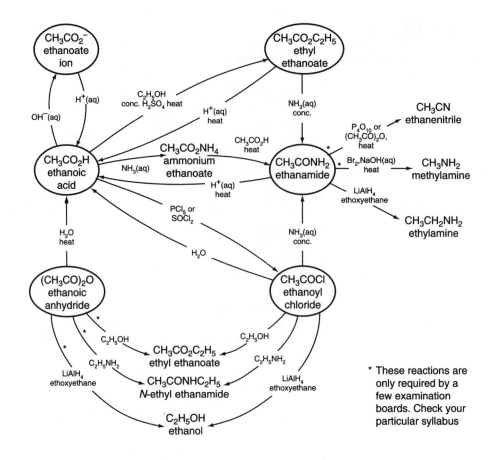

phenylmethylbenzoate

EXERCISE 30

a $C_3H_7CO_2H \xrightarrow{NH_3(aq)} C_3H_7CO_2^-NH_4^+ \xrightarrow[\text{excess } C_3H_7CO_2H]{\text{Heat with}} C_3H_7CONH_2$

$C_3H_7CONH_2 \xrightarrow[\text{Reflux}]{Br_2, OH^-(aq)} C_3H_7NH_2$

b $CH_3CHO \xrightarrow[\text{Reflux}]{Cr_2O_7^{2-}, H^+(aq)} CH_3CO_2H \xrightarrow{NH_3(aq)} CH_3CO_2^-NH_4^+ \xrightarrow[\substack{\text{excess} \\ CH_3CO_2H}]{\text{Heat with}} CH_3CONH_2$

c $CH_3CONH_2 \xrightarrow[\text{Heat, acidify}]{OH^-(aq)} CH_3CO_2H \xrightarrow[\text{Heat}]{C_2H_5OH, \text{ conc. } H_2SO_4} CH_3CO_2C_2H_5$

d $C_6H_5CO_2H \xrightarrow{SOCl_2*} C_6H_5COCl \xrightarrow{C_2H_5OH} C_6H_5CO_2C_2H_5$

e $CH_3CO_2H \xrightarrow[\text{Reflux}]{LiAlH_4, \text{ ether}} CH_3CH_2OH \xrightarrow{PCl_5} C_2H_5Cl \xrightarrow[\text{Reflux}]{KCN, \text{ ethanol}} C_2H_5CN$

$C_2H_5CN \xrightarrow[\text{Reflux, acidify}]{OH^-(aq)} C_2H_5CO_2H$

f $C_3H_7OH \xrightarrow[\text{Reflux}]{Cr_2O_7{}^{2-}, H^+(aq)} C_2H_5CO_2H \xrightarrow{PCl_5} C_2H_5COCl \xrightarrow{NH_3(aq)} C_2H_5CONH_2$

$C_2H_5CONH_2 \xrightarrow[\text{Reflux}]{Br_2, OH^-(aq)} C_2H_5NH_2 \xrightarrow[\text{Heat}]{NaNO_2, H^+(aq)} C_2H_5OH$

EXERCISE 31

$C_6H_6 \quad (1) \quad C_6H_5NO_2 \quad (2) \quad C_6H_5NH_2$
benzene $\longrightarrow$ nitrobenzene $\longrightarrow$ phenylamine

(5)
$\longrightarrow C_6H_5CONHC_6H_5$
N-phenylbenzamide

$C_6H_5CH_3 \quad (3) \quad C_6H_5CO_2H \quad (4) \quad C_6H_5COCl$
methylbenzene $\longrightarrow$ benzoic acid $\longrightarrow$ benzoyl chloride

1. Heat benzene at 60 °C with a mixture of concentrated nitric and sulphuric acids.
2. Reflux nitrobenzene with tin and concentrated hydrochloric acid and then neutralise the mixture with an alkali.
3. Reflux methylbenzene with acidified aqueous potassium manganate(VII).
4. Warm benzoic acid with phosphorus trichloride or sulphur dichloride oxide.*
5. Mix phenylamine and benzoyl chloride in the presence of an alkali.

EXERCISE 32

a i) **V** C_2H_5Br
   W CH_3CONH_2
   ii) Stage **A**: NH_3, ethanol.
       Stage **B**: $Cr_2O_7{}^{2-}$, $H^+(aq)$.
       Stage **C**: PCl_5 or $SOCl_2$.
       Stage **D**: P_4O_{10} or $(CH_3CO)_2O$.
       Stage **E**: $LiAlH_4$, ethoxyethane.
   iii) Stage **B** – oxidation.
b $CH_3CONHC_2H_5$

EXERCISE 33

a Amide, carboxyl, benzene ring (any two).
b
⬡—$COCl$ and $NH_2CH_2CO_2H$

c $H^+(aq)$, reflux.

* not PCl_5 as benzoic acid is also solid

EXERCISE 34 **a** **E** ethyl ethanoate, $CH_3CO_2C_2H_5$
F sodium ethanoate, $CH_3CO_2^-Na^+$
G ethanoyl chloride, CH_3COCl

b i) $CH_3CO_2H \xrightarrow{\text{NaOH(aq)}} CH_3CO_2Na$

ii) $CH_3CO_2H \xrightarrow[\text{Cold addition}]{\text{PCl}_5} CH_3COCl$

c i) H^+(aq), heat.

ii) $CH_3CO_2C_2H_5 + H_2O \xrightarrow[\text{Heat}]{H^+} CH_3CO_2H + C_2H_5OH$
  ethanoic acid ethanol

iii) Hydrolysis.
iv) Fractional distillation.

d i) $CH_3COCl + C_2H_5NH_2 \longrightarrow CH_3CONHC_2H_5 + HCl$
ii) *N*-substituted amide

EXERCISE 35 **a**

b

propane-1,2,3-triol $3CH_3CO_2H$

EXERCISE 36 **a** Oleic acid (two molecules) and linoleic acid.
b • Saturated fatty acids have no carbon–carbon double bonds. Examples of this type are lauric, myristic, palmitic and stearic acids.
  • Mono-unsaturated fatty acids have one carbon–carbon double bond. Examples of this type are palmitoleic, oleic and elaidic acids.
  • Polyunsaturated fatty acids have more than one carbon–carbon double bond. Examples are linoleic, linolenic and arachidonic acids.

EXERCISE 37 **a** Fats (triesters) with long-chain fatty acid side-chains have more electrons, and therefore have stronger intermolecular forces, and so higher melting points than those that are shorter.
b One or more carbon–carbon double bonds in the side-chain fatty acids with *cis* conformation cause the chain to kink. This reduces the intermolecular forces between the fat (triester) molecules because they cannot pack so tightly together. So unsaturated fats have lower melting points and are mostly liquid oils at room temperature.
c *Trans* fatty acids unlike *cis* fatty acids are not kinked. *Trans* fats will therefore pack more closely and have stronger intermolecular forces. This gives rise to higher melting points than in *cis* fats.

EXERCISE 38 a Hydrogenation of the unsaturated vegetable oil will take place by bubbling hydrogen through the oil at 180 °C in the presence of a nickel catalyst.

b

c The product of the reaction above will no longer have a carbon–carbon double bond. This means there will be no sharp bend in the side-chain fatty acids. Intermolecular forces will now be stronger between neighbouring fat molecules as they now pack together more closely. This will raise the melting point of the fat so it will be solid at room temperature.

EXERCISE 39 a

cis   *trans*

b In order to convert *cis* form into *trans* form the π bond must either be weakened or temporarily broken by the catalyst to enable a change into the *trans* position.

EXERCISE 40 a By keeping the mixture of oil and iodine monochloride in the dark, free radical substitution of alkyl groups is prevented.
b $ICl(aq) + KI(aq) \rightarrow KCl(aq) + I_2(aq)$
c Substituting into the expression:

$$c = \frac{n}{V} \text{ in the form } n = cV$$

gives $n = 0.100 \text{ mol dm}^{-3} \times 0.0400 \text{ dm}^3 = \textbf{0.00400 mol}$
d From the equation, 0.00400 mol $Na_2S_2O_3$ reacts with **0.00200 mol** liberated iodine.
   From the equation in (b): 0.00200 mol I_2 was liberated by **0.00200** mol of unreacted ICl.
e Total amount of ICl added to the oil can be obtained from the expression:

$$c = \frac{n}{V} \text{ in the form } n = cV$$

$$n = 0.100 \text{ mol dm}^{-3} \times 0.0250 \text{ dm}^3$$
$$= 0.00250 \text{ mol}$$
$$\therefore \text{ amount of ICl reacting with oil} = 0.00250 - 0.00200$$
$$= \textbf{0.000500 mol}$$

f 0.0005 mol ICl is equivalent to 0.0005 mol I_2. Substituting into the expression:

$$n = \frac{m}{M} \text{ in the form } m = nM$$

gives $m = 0.000500 \text{ mol} \times 254 \text{ g mol}^{-1}$
$= 0.127 \text{ g } I_2$

Since 0.127 g I_2 reacts with 0.127 g fat, then 100 g I_2 reacts with 100 g fat, therefore iodine number = **100**.

EXERCISE 41

a High cholesterol levels cause a build up of cholesterol in the artery walls, making them narrower and adversely affecting the blood supply to the heart.

b A high total cholesterol level may not necessarily be a bad thing since much of it may be HDL-cholesterol (good cholesterol).

c Eat less saturated fat, found mostly in animal fat apart from fish oils, and avoid a high intake of *trans* fats. Replace saturated fats with mono-unsaturated fats where possible.

d HDL-cholesterol acts like a sponge and rids the body of cholesterol.

EXERCISE 42

a

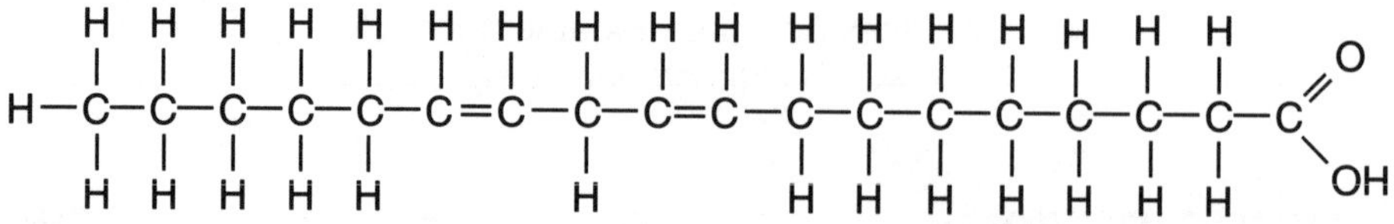

or any isomer as long as there are 18 carbon atoms and two carbon–carbon double bonds.

b i) Stearic acid is saturated because it has no carbon–carbon double bonds.
ii) Linoleic acid is unsaturated because it has carbon–carbon double bonds.

c

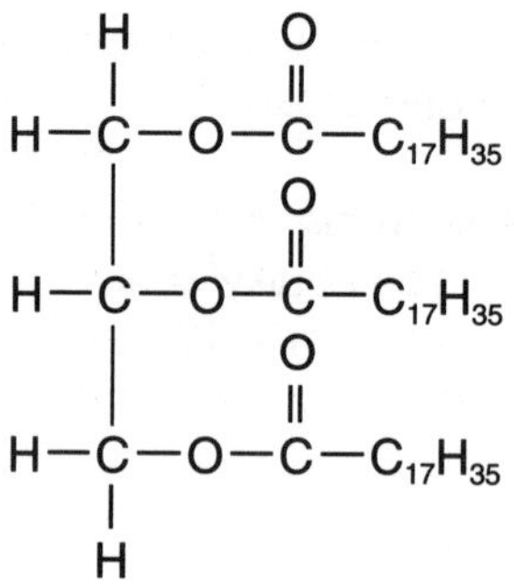

d i) Rapeseed oil.
ii) Rapeseed oil.
iii) Sunflower seed oil.
iv) Palm oil. There is a link between a diet high in saturated fats and increased risk of heart disease.

EXERCISE 43

$CH_2OCOC_{17}H_{35}$ | $CHOCOC_{17}H_{35}$ + 3NaOH | $CH_2OCOC_{17}H_{35}$	⟶	CH_2OH | $CHOH$ | CH_2OH	+	$3C_{17}H_{35}CO_2^-Na^+$
stearic acid (animal fat)		propane-1,2,3-triol (glycerol)		sodium stearate (soap)

EXERCISE 44

a Soaps are rather ineffective and wasteful in hard water because they react with dissolved calcium and magnesium ions to form precipitates (called 'scum') such as calcium stearate (calcium octadecanoate), $(C_{17}H_{35}CO_2^-)_2Ca^{2+}$. Soapless detergents do not form scums in hard water because the corresponding calcium and magnesium salts are soluble.

Soaps do not work in acidic conditions or in sea water. Hydrogen ions cause precipitation of the free acid, and sodium ions cause precipitation of the soap by the common ion effect. See ILPAC 6, Equilibrium I: Principles.

Soaps tend to be more expensive (for equivalent effects) because they are made from refined animal and vegetable fats whereas soapless detergents are made from the by-products of oil refining. (The comparison may change as oil becomes more expensive.)

b The hydrocarbon 'tails' of early detergents consisted of branched-chain polypropenes which are resistant to bacterial attack. They emerged unchanged from sewage works causing damage to wildlife and often giving rise to masses of dirty foam in rivers and on seashores. Most detergents are now biodegradable; the hydrocarbon 'tails' are either unbranched or have a single short branch and these are more readily broken down by bacteria.

EXERCISE 45   a

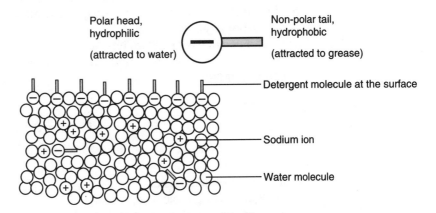

Detergent molecules at the surface reduce the surface tension of water because this arrangement breaks up the network of hydrogen bonding between water molecules. The water is then able to wet more thoroughly.

b

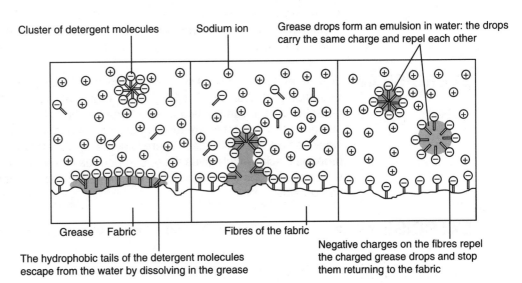

The mutual repulsion of the polar 'heads' forces the surface of grease into a spherical shape which can then leave the fabric.

EXERCISE 46 **a** (i), (iii) and (iv) are anionic detergents.
(ii) and (v) are cationic.
b Phosphates remove Mg^{2+} and Ca^{2+} ions. They combine with Mg^{2+} and Ca^{2+} ions which cause hardness.
c Increased levels of phosphates in waterways increase the population of algae making the water cloudy and scummy. The surface algae prevent light getting to the plants beneath the surface, which die.The death of these algae causes a drop in the oxygen content of the water which in turn deprives other organisms, such as fish, of oxygen. This process is known as eutrophication.
d Biological washing powders contain enzymes that would be denatured (structural changes) at high temperatures. These structural changes render the enzymes ineffective as catalysts.

EXERCISE 47 **a** All contain an amino group and a carboxyl group attached to the same carbon atom. (Proline has an —NH group instead of an —NH_2 group.)
b The letter α shows that the amino group is attached to the carbon atom **nearest** to the carboxyl group.
   (The use of Greek letters has been largely superseded by a numerical system, but note that α becomes 2 and not 1 because 1 refers to the carbon atom in the carboxyl group.)

$$\overset{\gamma}{\underset{4}{-C}}-\overset{\beta}{\underset{3}{C}}-\overset{\alpha}{\underset{2}{C}}-\underset{1}{CO_2H}$$

c i)   Aminoethanoic acid,
   ii) 2-aminopropanoic acid,
   iii) 2-amino-3-hydroxypropanoic acid,
   iv) 2-amino-3-phenylpropanoic acid,
   v) 2,6-diaminohexanoic acid,
   vi) 2-amino-3-methylbutanoic acid.
d Systematic names are not used because they are much longer and not simple to abbreviate.
e Glycine is the only amino acid of those listed that would not be optically active. It does not contain an asymmetric carbon atom and hence would not exist as enantiomers.

EXERCISE 48 Glycine exists in the solid state as an 'internal salt', consisting of doubly charged zwitterion particles, $NH_3^+CH_2CO_2^-$. The strong attraction of these particles for each other leads to a high melting point, characteristic of ionic compounds. The other three compounds in the table are all covalent liquids and their particles are held together by weak van der Waals' forces. There is some hydrogen bonding in 1-aminobutane and propanoic acid, which gives these compounds higher melting points than methyl ethanoate but the attractive forces are, nevertheless, far weaker than in ionic compounds.

EXERCISE 49 **a**

$$CH_3CHCO_2H(aq) + NaOH(aq) \longrightarrow CH_3CHCO_2^-Na^+(aq) + H_2O(l)$$
$$||$$
$$NH_2NH_2$$

b

$$CH_3CHCO_2H(aq) + HCl(aq) \longrightarrow CH_3CHCO_2H(aq)$$
$$||$$
$$NH_2NH_3^+Cl^-$$

EXERCISE 50   **a** Form **C** would predominate at low pH. There is a high concentration of hydrogen ions so that the equilibrium would move to the right (as written).
  Form **A** would predominate at high pH. The concentration of hydrogen ions is low which would cause the equilibrium to move to the left (as written).

b **B** is a zwitterion.
  A is the conjugate base of the zwitterion.
  C is the conjugate acid of the zwitterion.

c Form **B**, the zwitterion, would show no net movement in an electric field as it would be equally attracted to both electrodes.

EXERCISE 51   **a** Acidic: aspartic acid and glutamic acid (two acidic groups, one basic).
  Basic: arginine (four basic groups, one acidic).
  Basic: lysine (two basic groups, one acidic).
  Neutral: glycine and leucine (one basic group and one acidic).

b The isoelectric point tends to be low for acidic, high for basic and around the neutral point for neutral amino acids.

EXERCISE 52   **a** i) Proteins or amino acids whose isoelectric point is lower than the pH of the buffer will exist as anions, for example $H_2NCHRCO_2^-$ and move towards the anode.
  ii) Those amino acids or proteins whose isoelectric point is greater than the pH of the buffer will exist as cations, for example $H_3N^+CHRCO_2H$, and move towards the cathode.

b Any amino acids whose isoelectric point is the same as the buffer will not move since they will be in the dipolar form: $H_3N^+CHRCO_2^-$.

c The rate of migration of a particular amino acid or protein depends on its mass/charge ratio and the magnitude of the applied voltage.

EXERCISE 53   The correct sequence for the stages involved in the process of DNA fingerprinting is:

$$C\ E\ A\ G\ H\ F\ D\ B$$

EXERCISE 54   The boy's father is A. All the bands in the child's fingerprint also appear in either the mother's fingerprint or that of father A. B's fingerprint has very few bands in common with that of the child.

EXERCISE 55   **a** At **A**: $H_3N^+CH_2CO_2H$
  At **C**: $H_3N^+CH_2CO_2^-$
  At **E**: $H_2NCH_2CO_2^-$

b Point **B** on the curve is at pH 2.34, i.e. pH = pK_1. pK_1 refers to the equilibrium:

$$H_3N^+CH_2CO_2H(aq) \rightleftharpoons H_3N^+CH_2CO_2^-(aq) + H^+(aq)$$
$$\text{acid} \qquad\qquad\qquad \text{base}$$

and $pH = pK_1 - \log\dfrac{\text{[acid form]}}{\text{[base form]}}$

Since $pH = pK_1$, $\log\dfrac{\text{[acid form]}}{\text{[base form]}} = 0$

∴ [acid form] = [base form]

i.e. at point **B** there are equal concentrations of $H_3N^+CH_2CO_2H$ and $H_3N^+CH_2CO_2^-$.

Point **D** on the curve is at pH 10.60, i.e. pH = pK_2 + 1

pK_2 refers to the equilibrium:

$$H_3N^+CH_2CO_2^- \text{(aq)} \rightleftharpoons H_2NCH_2CO_2^- \text{(aq)} + H^+\text{(aq)}$$
$$\text{acid} \qquad\qquad\qquad \text{base}$$

and pH = pK_2 − log $\dfrac{\text{[acid form]}}{\text{[base form]}}$

Since pH = pK_2 + 1, log $\dfrac{\text{[acid form]}}{\text{[base form]}}$ = −1

∴ [base form] = 10[acid form]

i.e. at point **D**, $H_2NCH_2CO_2^-$ and $H_3N^+CH_2CO_2^-$ are present in the ratio 10 : 1.

c Glycine would act as a good buffer when sufficient acid or alkali has been added to convert half of it to the acidic or basic forms respectively, i.e. at pH = pK_1 and pK_2. The curve shows that near the points **B** and **D**, the pH changes least for the addition of a given amount of acid or alkali.

Glycine is a very poor buffer at pH values around its isoelectric point. Here, there is a very large change in pH for small additions of acid or alkali.

EXERCISE 56

a i) Alanine is a 'neutral' amino acid since it has one acidic group and one basic group. Consequently, the zwitterion, $CH_3CH(NH_3^+)CO_2^-$, predominates at pH = 7. When acid is added, this form accepts protons and is progressively converted into the acid form, $CH_3CH(NH_3^+)CO_2H$, the change being virtually complete when pH = 1.5.

$$CH_3CH(NH_3^+)CO_2^- \text{(aq)} + H^+\text{(aq)} \rightleftharpoons CH_3CH(NH_3^+)CO_2H\text{(aq)}$$

At a pH of 2.4, when pH = pK_1, the two forms are present in equal concentrations because:

$$\text{pH} = pK_1 - \log \frac{\text{[acid form]}}{\text{[base form]}}$$

(Here, the zwitterion behaves as a base.)

ii) When alkali is added to the neutral solution, the zwitterion gives up protons and is progressively converted into the basic form, $CH_3CH(NH_2)CO_2^-$, the change being virtually complete when pH = 11.5.

$$CH_3CH(NH_3^+)CO_2^- \text{(aq)} \rightleftharpoons CH_3CH(NH_2)CO_2^- \text{(aq)} + H^+\text{(aq)}$$

At a pH of 9.7, i.e. pH = pK_2, the basic form and the zwitterion are present in equal concentrations because:

$$\text{pH} = pK_2 - \log \frac{\text{[acid form]}}{\text{[base form]}}$$

(In this reaction, the zwitterion behaves as an acid.)

b Lysine has two basic groups and one acidic group. Consequently, there are three pK values, corresponding to the changes:

$NH_3^+(CH_2)_4CH(NH_3^+)CO_2H$	Acidic form	(Note that CO_2H loses a
$\pm\,H^+\,\updownarrow$	pK_1 = 2.2	proton more readily
$NH_3^+(CH_2)_4CH(NH_3^+)CO_2^-$	Zwitterion 1	than NH_3^+ and that the
$\pm\,H^+\,\updownarrow$	pK_2 = 9.0	NH_3^+ nearest the CO_2^-
$NH_3^+(CH_2)_4CH(NH_2)CO_2^-$	Zwitterion 2	loses a proton more
$\pm\,H^+\,\updownarrow$	pK_3 = 10.5	readily than the
$NH_2(CH_2)_4CH(NH_2)CO_2^-$	Basic form	farther NH_3^+ does.)

At a pH of 7, lysine is predominantly in the zwitterion 1 form. Addition of acid converts it progressively to the acidic form, the change being virtually complete when pH = 1.5. Addition of alkali converts zwitterion 1 to zwitterion 2 and then to the basic form, the change being virtually complete when pH = 11.5. By application of the equation:

$$pH = pK - \log \frac{[\text{acid form}]}{[\text{base form}]}$$

we see that the following relationships hold:
at pH = 2.2   [Acidic form] = [Zwitterion 1]
at pH = 9.0   [Zwitterion 1] = [Zwitterion 2]
at pH = 10.5   [Zwitterion 2] = [Basic form]

c The pK_a values for aspartic acid refer to the changes:

$CO_2HCH_2CH(NH_3^+)CO_2H$	Acidic form
$\pm\,H^+\,\updownarrow$	pK_1 = 2.1
$CO_2^-CH_2CH(NH_3^+)CO_2H$	Zwitterion 1
$\pm\,H^+\,\updownarrow$	pK_2 = 3.9
$CO_2^-CH_2CH(NH_3^+)CO_2^-$	Zwitterion 2
$\pm\,H^+\,\updownarrow$	pK_3 = 9.8
$CO_2^-CH_2CH(NH_2)CO_2^-$	Basic form

At pH = 1, the acidic form, $CO_2HCH_2CH(NH_3^+)CO_2H$, predominates.
At pH = 7, the zwitterion, $CO_2^-CH_2CH(NH_3^+)CO_2^-$, predominates.
At pH = 13, the basic form, $CO_2^-CH_2CH(NH_2)CO_2^-$, predominates.

EXERCISE 57   **a**

$$CH_3\underset{\underset{NH_2}{|}}{C}HCO_2H + HONO \longrightarrow CH_3\underset{\underset{OH}{|}}{C}HCO_2H + N_2 + H_2O$$

b i) An acyl chloride, e.g. ethanoyl chloride, CH_3COCl, or a carboxylic acid anhydride, e.g. ethanoic anhydride, $(CH_3CO)_2O$, could be used to acylate glycine.

   ii)

$$CH_3\underset{\underset{NH_2}{|}}{C}HCO_2H + CH_3COCl \longrightarrow CH_3\underset{\underset{NHCOCH_3}{|}}{C}HCO_2H + HCl$$

c i)

$$CH_3\underset{\underset{NH_2}{|}}{C}HCO_2H + C_2H_5OH \longrightarrow CH_3\underset{\underset{NH_2}{|}}{C}HCO_2C_2H_5 + H_2O$$

   ii) Dry hydrogen chloride is a suitable catalyst (giving the salt form of the product, $RNH_3^+Cl^-$).

EXERCISE 58 G = glycine, $CH_2(NH_2)CO_2H$ (or $CH_2(NH_3^+)CO_2^-$)
H = methylamine, CH_3NH_2
I = hydroxyethanoic acid, CH_2OHCO_2H
The high melting point of G and its solubility in water suggests an ionic compound. The formation of crystalline salts with both acids and bases suggests an amino acid. This is partly confirmed by decarboxylation with soda-lime yielding H, which could be an amine since simple amines burn readily in air and dissolve in water to give alkalis. Further confirmation is given by the reaction of nitrous acid with G, which replaces the —NH_2 group by an —OH group and releases nitrogen and water.

If the amino acid is represented by $R(NH_2)CO_2H$, the compound I is $R(OH)CO_2H$. The relative molecular mass of R is therefore $76 - (48 + 12 + 2) = 14$, which corresponds to the formula CH_2. The equations for the reactions are as follows.
$$CH_2(NH_2)CO_2H(aq) + H^+(aq) \rightarrow CH_2(NH_3^+)CO_2H(aq)$$
$$CH_2(NH_2)CO_2H(aq) + OH^-(aq) \rightarrow CH_2(NH_2)CO_2^-(aq) + H_2O(l)$$
$$CH_2(NH_2)CO_2H(s) + 2NaOH(s) \rightarrow CH_3NH_2(g) + H_2O(l) + Na_2CO_3(s)$$
$$2CH_3NH_2(g) + 4\tfrac{1}{2}O_2(g) \rightarrow N_2(g) + 2CO_2(g) + 5H_2O(g)$$
$$CH_3NH_2(g) + H_2O(l) \rightarrow CH_3NH_3^+(aq) + OH^-(aq)$$
$$CH_2(NH_2)CO_2H(aq) + HONO(aq) \rightarrow CH_2(OH)CO_2H(aq) + N_2(g) + H_2O(l)$$

EXPERIMENT 5
Questions

1. The solution would be slightly alkaline. Since CO_3^{2-} is the conjugate base of the weak acid HCO_3^- which is, in turn, the conjugate base of the weak acid H_2CO_3, the following equilibria would be set up:

$$CO_3^{2-}(aq) + H^+(aq) \rightleftharpoons HCO_3^-(aq)$$
$$HCO_3^-(aq) + H^+(aq) \rightleftharpoons H_2CO_3(aq)$$

These would decrease the concentration of hydrogen ions in solution, making it slightly alkaline.
2. In alkaline solution, the zwitterion form of glycine would tend to lose a proton and become the basic form:

$$CH_3NH_3^+CO_2^-(aq) + OH^-(aq) \rightleftharpoons CH_2NH_2CO_2^-(aq) + H_2O(l)$$

3. The basic form of glycine can make two dative bonds, using the lone pair of electrons on the nitrogen and the negative charge on the carboxyl oxygen.

Glycine is therefore a bidentate ligand.

EXERCISE 59 a

(Note that this complex is sometimes called a salt because it arises from the combination of two ions.)
b This structure would be essentially square planar and would have no net charge.

EXERCISE 60

a

$$CH_3CO_2H \xrightarrow[\text{UV light}]{Cl_2} CH_2ClCO_2H \xrightarrow[\text{conc. (aq)}]{NH_3} CH_2NH_2CO_2^- NH_4^+ \xrightarrow{H^+(aq)} CH_2NH_2CO_2H$$

b

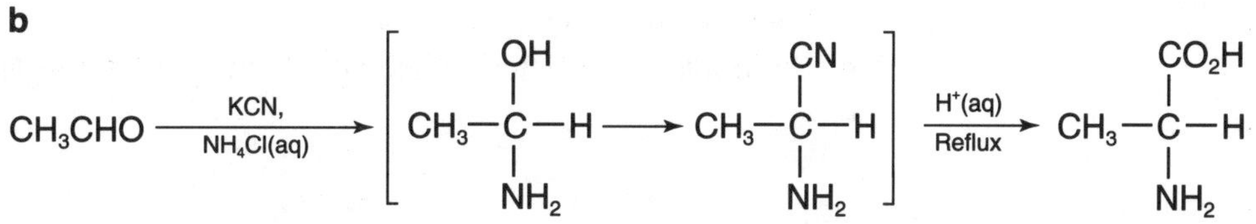

c Glycine is not optically active since its molecule has a plane of symmetry – there is no asymmetric carbon atom. Alanine is optically active because of the asymmetric carbon atom. However, in the reaction shown in (b), both enantiomers are formed to give a racemic mixture which is not active.

EXERCISE 61

a

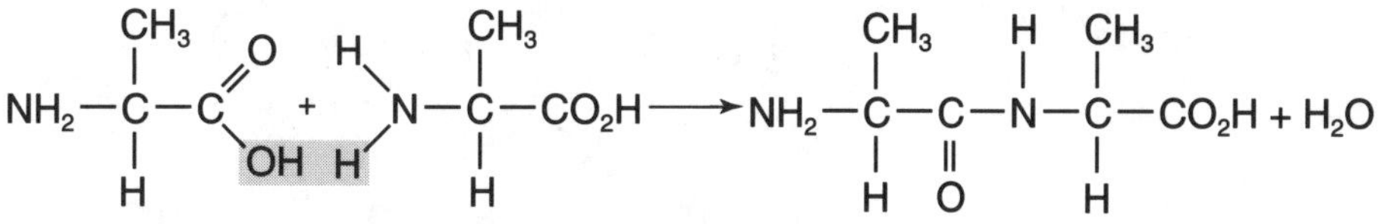

b In aqueous solution, alanine exists primarily as a zwitterion in which the functional groups are not —NH_2 and —CO_2H but —NH_3^+ and —CO_2^-. These groups do not undergo condensation reactions. Even if —NH_2 and —CO_2H groups were present, they would be more likely to react by proton transfer (to give —NH_3^+ and —CO_2^-) than by condensation.

c If the —CO_2H group in alanine is converted to a —$COCl$ group by reaction with phosphorus pentachloride, PCl_5, or sulphur oxide chloride, $SOCl_2$, a zwitterion cannot be formed by proton transfer. A condensation reaction readily occurs between the acyl chloride and another alanine molecule.

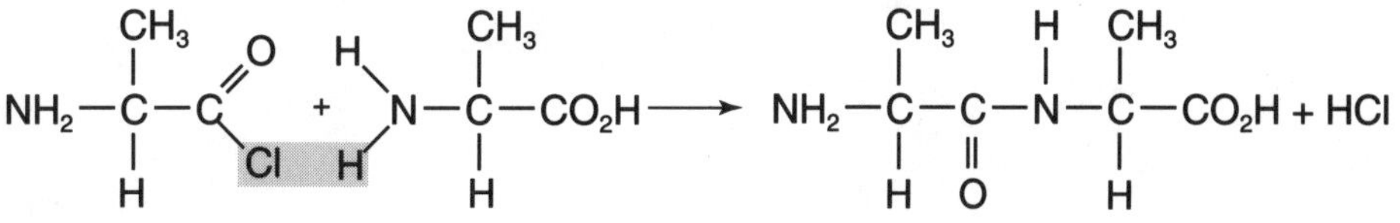

(Long chains will be formed as well. If the dimer **alone** is required, it is necessary to protect the —NH_2 group in half the alanine by acylation before introducing the —$COCl$ group.)

d The systematic name is too long and clumsy for regular use and the traditional name is well established. Moreover, the traditional name suggests how it might be made. Alanylalanine is an example of a dipeptide.

EXERCISE 62 **a** The central C—N bond is 0.132 nm long which is intermediate between the lengths of single and double bonds. This suggests that the bond has some double bond character due to overlap between the p orbitals of the lone pair on the N atom and those of the C = O double bond.

b There will be restricted rotation about the central C—N bond.

c

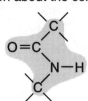

EXERCISE 63 **a**

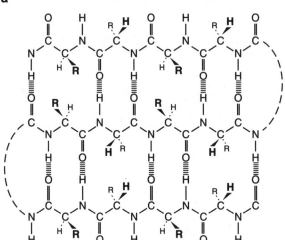 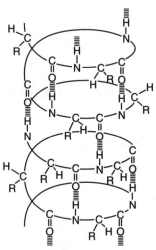

β-pleated-sheet structure of a protein   α-helix structure of a protein

b The R groups project outwards from the peptide links, roughly at right angles to the main axis of the chain.

EXERCISE 64 **a** α-Helical structure appears in the following ranges of residue numbers: 7–14, 25–35, 80–85, 91–99, 108–114, 119–125

b i) Cysteine.

ii)

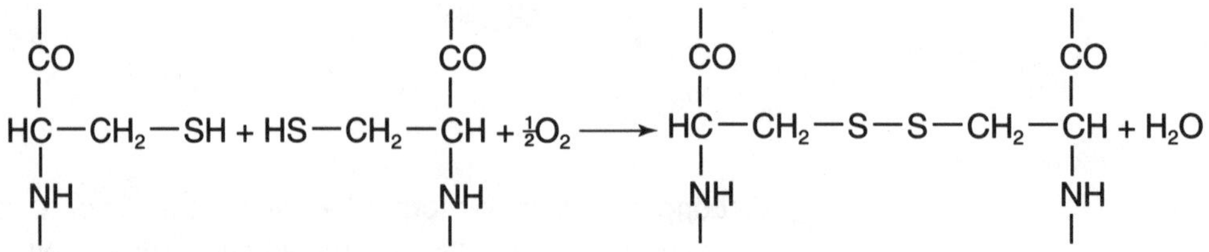

$$\text{HC}-\text{CH}_2-\text{SH} + \text{HS}-\text{CH}_2-\text{CH} + \tfrac{1}{2}\text{O}_2 \longrightarrow \text{HC}-\text{CH}_2-\text{S}-\text{S}-\text{CH}_2-\text{CH} + \text{H}_2\text{O}$$

iii) The conversion of cysteine to cystine is an oxidation reaction.

c Ionic bonding.

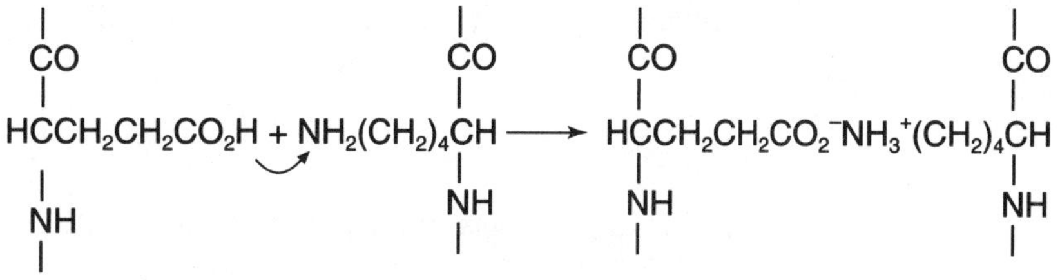

d In the denaturing (denaturation) of a protein such as lysozyme, the secondary and tertiary structures are destroyed, leaving the primary structure unaffected. Denaturing may be achieved in a number of ways, e.g. by heating above 70 °C, by the action of strong acids or alkalis and by dissolving in some organic solvents.

EXPERIMENT 6

Questions

1. The intensity of the purple (or pink) colour is related to the concentration of the protein. A colorimeter or spectrophotometer could be calibrated to enable concentration to be determined from colour intensity.

2.

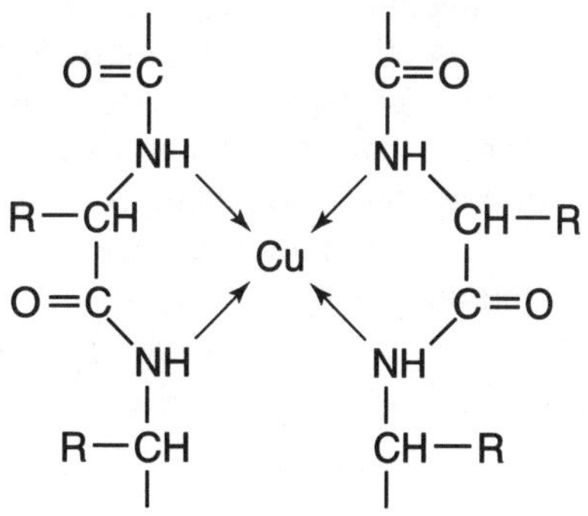

3. No, because the biuret complex is formed from nitrogen in peptide groups. However, care is needed because some amino acids form bluish complexes with copper(II) ions (see Experiment 5) and the colours can be confused.

EXERCISE 65

a

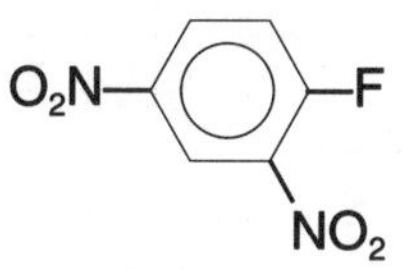

b

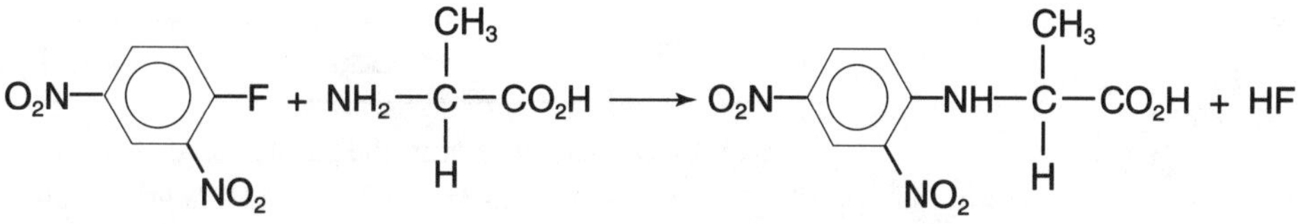

2,4-dinitrophenylalanine

c The hydrolysis-resistant bond allows the N-terminal end of the peptide chain to be identified. For instance, if alanine were the N-terminal amino acid, treatment with FDNB followed by hydrolysis would yield a yellow compound, 2,4-dinitrophenyl-alanine (DNP-ala). By chromatography, this can be separated from the other amino acids and identified.

EXERCISE 66   Item (D) shows that lysine is the N-terminal amino acid, because FDNB reacts selectively with terminal —NH₂ groups. Item (B) shows that lysine is linked to valine, valine to glycine, glycine to glycine and, finally, glycine to arginine:

$$\text{N} \qquad\qquad \text{C}$$
$$\text{lys–val–gly–gly–arg}$$

Confirmation of this order is given by item (C). The amount of pentapeptide is given by:

$$n = \frac{m}{M} = \frac{1.00\ \text{g}}{501\ \text{g mol}^{-1}} = 2.00 \times 10^{-3}\ \text{mol}$$

Item (A) shows that carboxypeptidase releases amino acids progressively from the C-terminal end, giving an equal amount of arginine and twice as much glycine.

EXERCISE 67   The peptide is first hydrolysed by refluxing with 6 M HCl for about 20 hours. Excess acid is removed by evaporation on a steam bath and the solid residue is dissolved in a little warm water. The amino acids are separated and identified by chromatography. (You would be expected to give an account of the technique after you have completed the next section.) The separate spots of amino acids can be re-dissolved from the chromatography paper (or thin-layer adsorbant), evaporated to dryness and weighed to give their relative amounts.

The structure of a dipeptide can be determined by means of its reaction with 1-fluoro-2,4-dinitrobenzene (FDNB), which combines with the terminal —NH₂ group. Subsequent hydrolysis yields one amino acid and one dinitrophenyl amino acid, and these can be identified by chromatography to give the structure of the dipeptide.

The two dipeptides are gly–ala and ala–ala. Since there is no ala–gly, the sequence must be:

$$\text{N} \qquad\qquad \text{C}$$
$$\text{gly–ala–ala–ala}$$

EXERCISE 68   a

Table 14

Type of chromatography	Separation phases		Principle* – adsorption or partition
	Mobile	**Stationary**	
Column	Liquid	Solid	Adsorption†
Thin-layer	Liquid	Solid	Adsorption
Paper	Liquid	Liquid	Partition
Gas–liquid	Gas	Liquid	Partition

* In some cases, **both** adsorption and partition occur, but one is nearly always predominant.
† Partition between eluent and adsorbed liquid may be important in some cases.

b In thin-layer chromatography, a solid such as silica gel, alumina or celite is made into a slurry with water. Glass or plastic plates are dipped into the slurry and dried to give a thin firm layer of solid.

Column chromatography can be carried out with almost any finely ground, inert solid. Commonly used are silica gel, charcoal, alumina, magnesium or sodium carbonate, starch, talc or sucrose.

c i) The 'activation' of chromatographic materials, such as charcoal, involves the removal of adsorbed substances by means of strong heating.

ii) Activation promotes adsorption in chromatography so that separation of substances is more readily achieved.

d The rate at which a substance moves during chromatography depends on the relative importance of two factors:

i) its solubility in the mobile phase,

ii) the extent of its adsorption by (or solubility in) the stationary phase.

The fastest moving substances are those that are very soluble in the mobile phase and weakly adsorbed by (or sparingly soluble in) the stationary phase. The opposite properties cause slow movement.

EXPERIMENT 7

Specimen results

Results Table 5

Amino acid	Distances travelled/cm		R_f value
	By solvent	By amino acid	
Aspartic acid – alone	7.1	1.1	0.15
– in mixture	6.8	0.9	0.13
Leucine – alone	7.0	5.5	0.79
– in mixture	6.8	5.1	0.75
Lysine – alone	7.1	2.6	0.37
– in mixture	6.8	2.4	0.35

Questions

1. R_f values vary slightly with local conditions such as temperature, the purity of the solvent, the nature of the chromatography paper, air pressure and humidity.

2. The greater the solubility of a substance in a particular solvent, the less readily it is retained by the stationary phase and the faster it appears to move. This gives a higher R_f value.

3. Amino acids are readily transferred from moist and/or dirty fingers on to the paper and show up as confusing brown marks when sprayed with ninhydrin solution.

EXERCISE 69

a Amino acid **A** R_f value in phenol $= \dfrac{23.5 \text{ mm}}{53.5 \text{ mm}} = \mathbf{0.44}$

R_f value in butan-1-ol/ethanoic acid $= \dfrac{20.0 \text{ mm}}{53.5 \text{ mm}} = \mathbf{0.37}$

∴ **A** is **alanine**

Amino acid **B** R_f value in phenol $= \dfrac{35.0 \text{ mm}}{53.5 \text{ mm}} = \mathbf{0.65}$

R_f value in butan-1-ol/ethanoic acid $= \dfrac{36.0 \text{ mm}}{53.5 \text{ mm}} = \mathbf{0.67}$

∴ **B** is **phenylalanine**

b Glycine must be the N-terminal amino acid.

c Amount of polypeptide $= \dfrac{1.00 \text{ g}}{780 \text{ g mol}^{-1}} = 1.28 \times 10^{-3} \text{ mol}$

Approximately the same amount of aspartic acid and **A** were produced together with twice as much serine. Since carboxypeptidase removes amino acids successively from the C-terminal end of the polypeptide, this shows that the four residues at that end are, in any order: asp, ser, ser, **A**.

d The most likely sequence is (iii).
  (i) has two aspartic acid residues and only one serine in the C-terminal quartet.
  (ii) has **B** instead of **A** in the C-terminal quartet.
  (iv) and (v) have asp, ser, ser and **A** at the N-terminal end.

EXERCISE 70   1. E   2. D   3. B   4. D   5. C

EXERCISE 71   Ethene-based addition polymers:

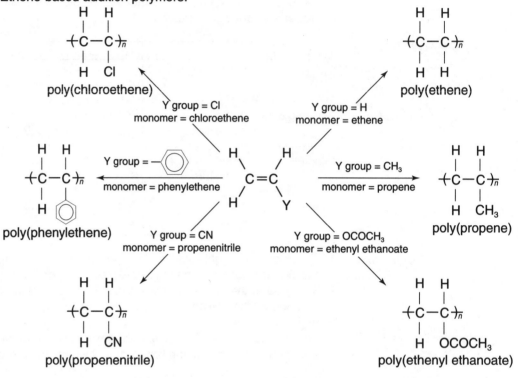

EXERCISE 72

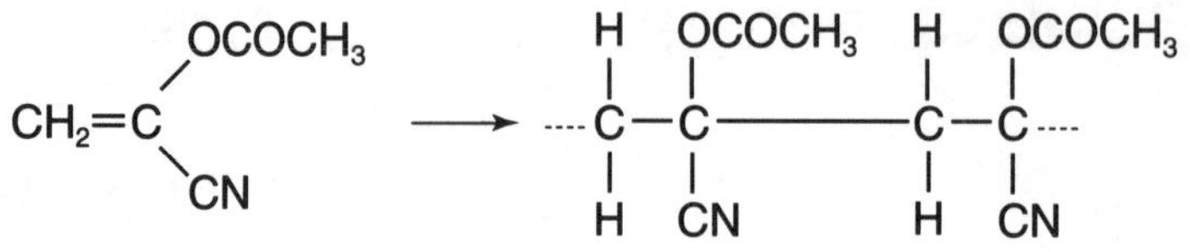

EXERCISE 73   **A** is isotactic (regular arrangement). **B** is atactic (irregular arrangement).

EXERCISE 74   **a**

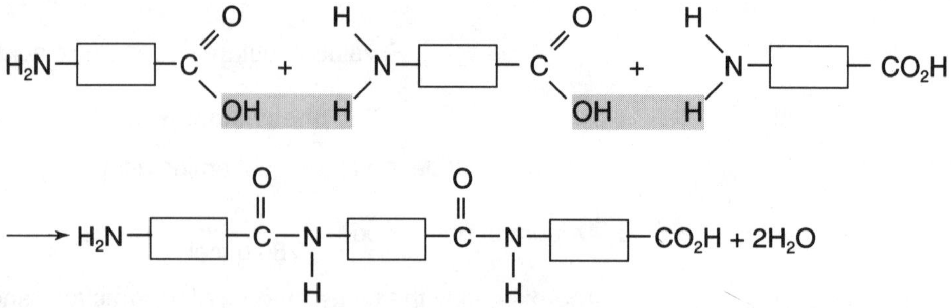

b The —CONH— link is an amide link (or bond). This is the same as the link between
amino acids in proteins, but there the term peptide link (or bond) is used.

EXERCISE 75 **a** HO_2C—☐—CO_2H + H_2N—☐—NH_2 →

HO_2C—☐—$CONH$—☐—NH_2 + H_2O

b

$$-\overset{\overset{\displaystyle O}{\|}}{C}-☐-\overset{\overset{\displaystyle O}{\|}}{C}-\underset{\underset{\displaystyle H}{|}}{N}-☐-\underset{\underset{\displaystyle H}{|}}{N}-$$

c A homopolymer, such as the one in Exercise 74, is made from a single monomer. A copolymer, such as the one in this exercise, is made from two different monomers which alternate in the final structure.

Note that a copolymer has a single repeating unit in its structure (see (b) above) but this unit is itself made from two monomers.

EXERCISE 76 **a** $-(CH_2)_5CONH-$ or $-\overset{\overset{\displaystyle }{}}{\underset{\underset{\displaystyle O}{\|}}{C}}-(CH_2)_5-\underset{\underset{\displaystyle H}{|}}{N}-$ or $-NH(CH_2)_5CO-$

b This type is called nylon-6 because the repeating unit has six carbon atoms.
c i) Both are polyamides. The monomers are similar in that they contain an NH_2 and a CO_2H group.
ii) Proteins are made up of a selection of up to twenty amino acid monomers, nylon-6 is made from just one. In other words, nylon-6 is a homopolymer, proteins are heteropolymers.

EXERCISE 77 **A** could be made from monomer 2 – homopolymer.
B could be made from monomers 1 and 3 – copolymer.
C could be made from monomer 4 – homopolymer.

EXERCISE 78 E.

EXERCISE 79 **a** $HOCH_2CH_2OH$ + HO_2C—⬡—CO_2H + $HOCH_2CH_2OH$

$$\longrightarrow 2H_2O + HOCH_2CH_2-O-\overset{\overset{\displaystyle O}{\|}}{C}-⬡-\overset{\overset{\displaystyle O}{\|}}{C}-O-CH_2CH_2OH$$

b Two ester linkages are formed by condensation between —OH and —CO_2H groups.
c Terylene (polyethylene terephthalate) is formed by repeated condensation reactions.

EXERCISE 80 1 and 3, but not 2 and 4.

EXERCISE 81 **a**

⬡—OH + CH_2O ⟶ ⬡(OH)(CH₂OH)

Similarly in the 3 and 5 positions.
b Electrophilic attack. The benzene ring in phenol is particularly susceptible to electrophilic attack because of the activating effect of the —OH group. The carbon atom in methanal has a partial positive charge because it is attached to a highly electronegative oxygen atom; this is a factor in making methanal an electrophile.

c

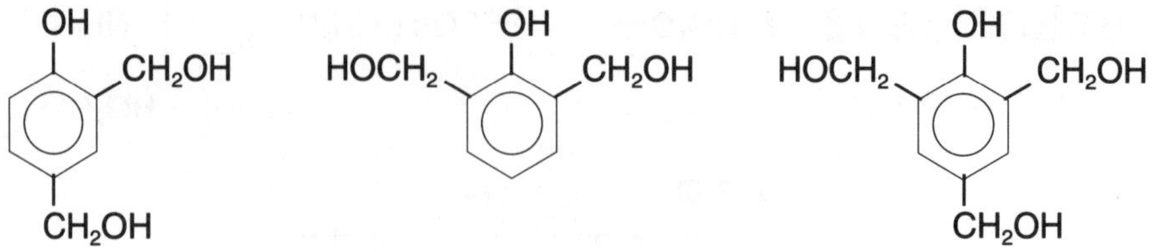

EXERCISE 82

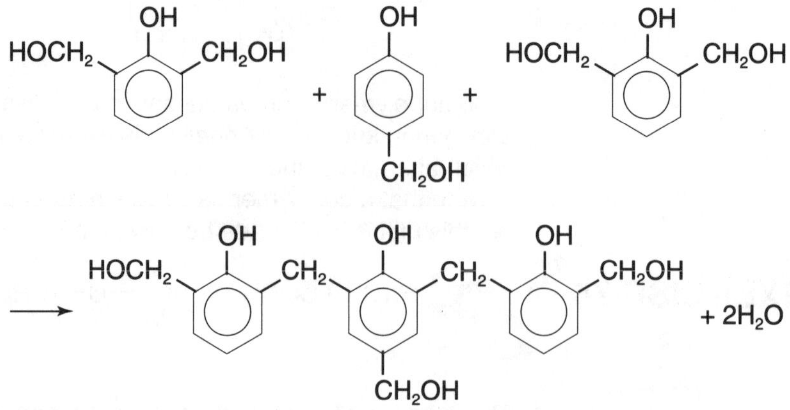

EXERCISE 83

Bakelite, once it has been extensively cross-linked, consists of a rigid three-dimensional giant structure in which all the bonds are fairly strong. Hardly any deformation can occur without breaking some of these bonds, consequently, the substance is hard and brittle. For the same reason, solvent molecules cannot penetrate the structure and surround discrete molecules.

EXERCISE 84

a X-ray diffraction reveals the crystalline nature of some polymers.

b i) A regular 'head-to-tail' arrangement of monomer units favours the formation of crystallites (but other factors may override this one).

ii) Atactic polymers have a random arrangement of attached groups so that regular crystalline packing is unlikely.

iii) Large groups attached to a polymer chain make regular packing difficult. Consequently, crystalline character is low.

iv) Extensive chain-branching makes regular packing difficult and is therefore associated with low crystalline character.

v) Extensive cross-linking is usually in random directions so that regular packing of polymer chains is unlikely and crystalline character is low.

vi) Thermosets have extensive cross-linking and, therefore, little crystalline character.

c Long unbranched poly(ethene) chains are able to pack together fairly closely to give a high degree of crystallinity. The closer packing results in higher density, higher melting point, greater strength and more rigidity.

EXERCISE 85

a A is a thermoplastic because it has no cross-links between the chains. B is a thermoset because it has cross-links between the chains.

b The term thermoplastic is applied to materials that soften and flow on the application of heat. These polymers have no cross-links between the chains. This means that on heating, molecular vibrations are able to overcome intermolecular forces and the chains move apart. They can therefore be moulded by heating them, shaping them and allowing them to cool.

c Thermosets have cross-links (covalent bonds) between the chains. This means they do not usually soften on heating. These cross-links also effectively prohibit chain movement so these polymers are brittle.

d Bakelite, which you studied in Exercises 82 and 83, is an example of a thermoset (others include carbamide methanal, melamine-methanal, polyurethanes, etc.).

EXERCISE 86 a

$$
\begin{array}{c}
\quad\;\; \text{H}\;\;\; \text{CH}_3 \qquad\qquad \text{H}\;\;\; \text{CH}_3 \\
\quad\;\; | \quad\;\; | \qquad\qquad\;\;\, | \quad\;\; | \\
-\text{C}-\text{C}\text{——}\text{C}-\text{C}- \\
\quad\;\; | \quad\;\; | \qquad\qquad\;\;\, | \quad\;\; | \\
\quad\;\; \text{H}\;\;\; \text{CO}_2\text{CH}_3 \quad\;\; \text{H}\;\;\; \text{CO}_2\text{CH}_3
\end{array}
$$

Perspex

b Perspex is brittle because the bulky side groups will hinder the movement of the polymer chains. Also the $C = O$ and $C - O$ groups are polar, so dipole–dipole as well as van der Waals' forces will arise between the side groups of neighbouring polymer chains, again restricting movement.

EXERCISE 87 When a sample of elastomer is stretched, there is some straightening of the long-chain molecules of the polymer. Since the favoured state for a long molecule is to be twisted and folded, the molecular chains return to this state when the stress is removed. Complete recovery from considerable stretching occurs because of the presence of a small number of cross-links which prevent the molecular chains from slipping past each other.

EXERCISE 88 a

$$
\begin{array}{c}
\text{CH}_2\text{=}\text{C}-\text{CH}\text{=}\text{CH}_2 \\
\quad\;\; | \\
\quad\;\; \text{CH}_3
\end{array}
$$

b

$$
\begin{array}{c}
-\text{CH}_2-\text{C}\text{=}\text{CH}-\text{CH}_2- \\
\quad\quad\; | \\
\quad\quad\; \text{CH}_3
\end{array}
$$

repeated indefinitely in a long chain

c Natural rubber is an addition polymer and a homopolymer.

d The element sulphur can be used in a hot vulcanisation process but a cold process using disulphur dichloride, S_2Cl_2, is more often employed.

e In vulcanisation, neighbouring chains become linked together at intervals by 'bridges' consisting of two or more sulphur atoms.

f The presence of double bonds allows addition reactions to occur so that links can be made between neighbouring chains.

g A small degree of cross-linking makes the rubber elastic, i.e. the chains can move relative to one another when pulled, but return to their original positions. As the extent of cross-linking increases, the rubber becomes stiffer until, at 30–50% sulphur content, the structure becomes rigid and hard (ebonite and vulcanite).

EXERCISE 89 a Ester groups are present in both DEMA and ATBC.

b The $C = O$ group in fats and plasticisers is polar so there will be dipole–dipole attraction between the two types of molecule.

c The $C - Cl$ bond in PVC will be polar so there will be dipole–dipole attraction between the chains. This will restrict movement between the chains to a certain extent, making PVC more brittle than poly(ethene), which has no polar groups.

d A plasticiser like ATBC has polar bonds and will get between the polymer chains so reducing the attraction between them.

e

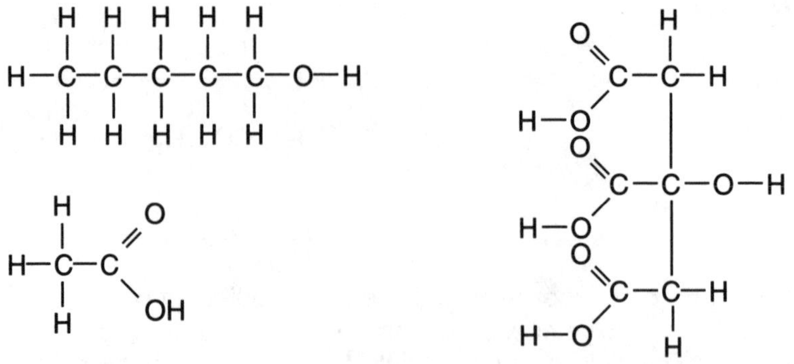

H   H   H   H   H
| | | | |
H—C—C—C—C—C—O—H
| | | | |
H   H   H   H   H

H
|
H—C—C
| ‖O
H   \OH

O   H
‖ |
C—C—H
H—O |
    O
    ‖
    C—C—O—H
H—O |
    O
    ‖
    C—C—H
H—O |
        H

EXERCISE 90   Both poly(propenenitrile) (in its isotactic form), nylon and polyesters have polymer chains which can be readily packed together in crystallites. When the material is drawn into a fibre, these crystalline areas become aligned with the axis of the fibre like logs floating down a river. Parallel fibres are then held together by hydrogen bonding (in nylon) or dipole attraction (in poly(propenenitrile) and polyesters).

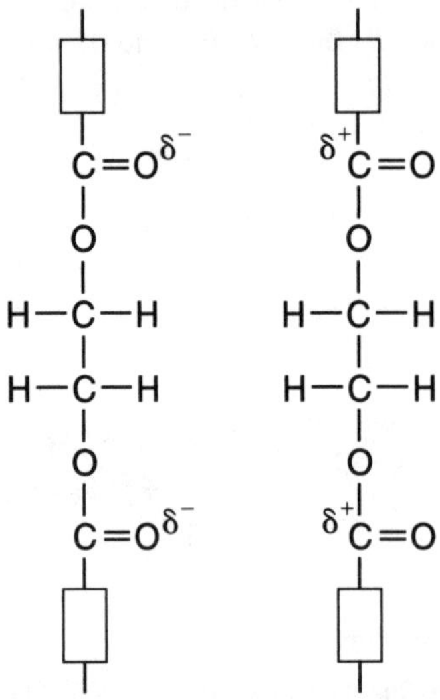

C=O$^{\delta^-}$ $^{\delta^+}$C=O
| |
O   O
| |
H—C—H   H—C—H
| |
H—C—H   H—C—H
| |
O   O
| |
C=O$^{\delta^-}$ $^{\delta^+}$C=O

polyester

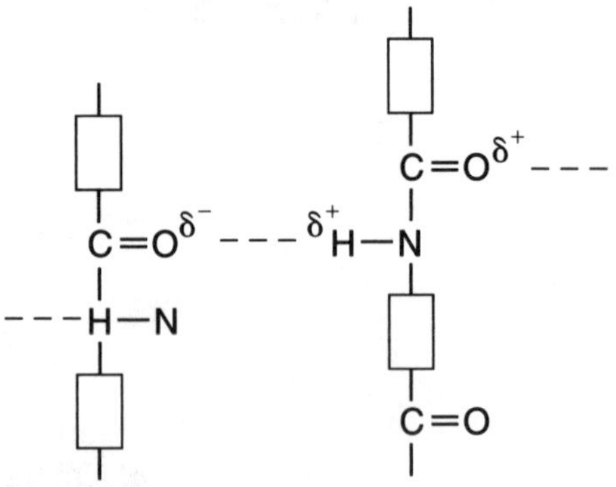

C=O$^{\delta^-}$ - - - $^{\delta^+}$H—N

- - -H—N

C=O$^{\delta^+}$ - - -

H—N

C=O

nylon

H—C—H   H—C—H
| |
H—C—CN$^{\delta^-}$ $^{\delta^+}$H—C—CN
| |
H—C—H   H—C—H
| |
H—C—CN$^{\delta^-}$ $^{\delta^+}$H—C—CN
| |
H—C—H   H—C—H

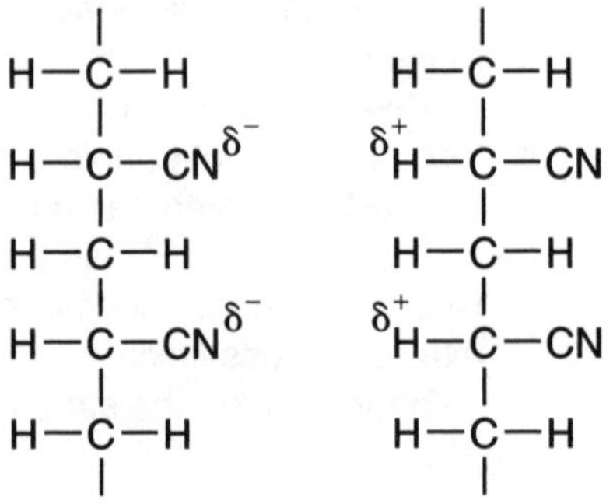

poly (propenenitrile)
(orlon)

Poly(phenylethene) is usually made by a method that results in an isotactic structure. This, coupled with the fact that the attached phenyl groups are bulky, makes the close alignment of polymer chains in fibres difficult to achieve. Consequently, fibres tend to be very weak.

Poly(ethene) can form fibres in which polymer chains are closely aligned but the bonding forces between chains are too weak to form a useful fibre.

EXERCISE 91   **a** **Low-density poly(ethene)**
In the original process ethene is polymerised in the liquid state at pressures up to 300 atm and temperatures up to 300 °C. Peroxides which yield free radicals are used to initiate the reaction.

$$nCH_2 = CH_2 \xrightarrow[\substack{\sim300\ °C \\ \sim300\ atm}]{\text{peroxides}} \ \ (CH_2-CH_2)_n$$
low-density
poly(ethene)

High-density poly(ethene)
This process was developed in the 1950s by Karl Ziegler using low pressure (<50 atm) and a temperature (<100 °C) and Ziegler–Natta type catalysts (organo-metallic complexes, often based on chromium or vanadium).

$$nCH_2 = CH_2 \xrightarrow[\substack{<50\ atm\ <100\ °C}]{\text{Ziegler–Natta catalyst}} \ \ (CH_2-CH_2)_n$$
high-density poly(ethene)

b Low-density poly(ethene) has long branches along its chain whereas high-density poly(ethene) has very few short branches along its chain. This means that the chains cannot pack as closely in low-density poly(ethene) as they can in the higher density form. Low-density poly(ethene) consequently will have a lower melting point and be more flexible.

c The manufacture of low-density poly(ethene) is by free radical mechanism.

d i–iv) Low-density poly(ethene) is more flexible and cheaper and will therefore be used to make plastic carrier bags (non-crinkly) and detergent squeeze bottles. High-density poly(ethene), being more rigid, will be used to make bleach bottles. The higher melting point of this form means it can be used in medical equipment that needs to be sterilised at high temperatures.

EXERCISE 92   **a** i) Alkylation (Friedel–Crafts reaction).
   ii) Dehydrogenation.

   b i) $AlCl_3$ as a catalyst, 90–100 °C.
   ii) ZnO or Fe_2O_3 as a catalyst, 600–630 °C.

   c i) A peroxide such as di(benzenecarbonyl) peroxide (benzoyl peroxide), $C_6H_5CO-O-O-COC_6H_5$.
   ii)

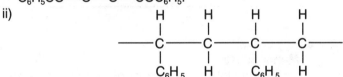

EXERCISE 93
Table 18

SPI code	Systematic name	Common name	Uses
1 PETE	Poly(ethenyl-1,4-benzene-dicarboxylate)	Polyethylene terephthalate (PET)	Video and audio tapes, garments, carpets, bottles, jars, floppy discs, ovenable cookware
2 HDPE	Poly(ethene) High-density	Polythene	Ice cream tubs, crinkly grocery bags, inner packaging for cereals, bleach bottles, buckets
3 V	Poly(chloroethene)	Polyvinylchloride (PVC)	Pipes, guttering, curtain rails, window frames, wall cladding, freezer bags, display boards, sewer pipes, fruit juice and vegetable oil bottles
4 LDPE	Poly(ethene) Low-density	Polythene	Garbage bags and bins, sacks, pond liners, stretch wrap film, squeeze bottles
5 PP	Poly(propene)	Polypropylene	Crisp and sweet wrappers, car bumpers, tough fibres, ropes, carpets, woven sacks, garden tables and chairs
6 PS	Poly(phenylethene)	Polystyrene	Containers for dairy products, disposable cups, beakers, cutlery, DIY packs, house foam insulation, toys, combs, plant pots, moulded foam protective packaging for fragile items

EXERCISE 94

a Key points:

Manufacture

1. Fermentation of a mixture of carbohydrates and organic acids . . .
2. . . . using the bacterium *Alcaligenes eutrophus*.
3. Cells harvested at the end of the growth cycle and broken open.
4. Polymer removed by solvent extraction and then purified.
5. Processing carried out between 185 °C and 190 °C.

Properties

6. Durability, stability and water resistance of conventional plastics.
7. Rapidly broken down above 205 °C.
8. Quickly (or efficiently) broken down into carbon dioxide and water . . .
9. . . . by fungi and bacteria.
10. Ester linkages hydrolysed and then metabolised.

Uses

11. In medical implants for slow-release capsules.

b i)

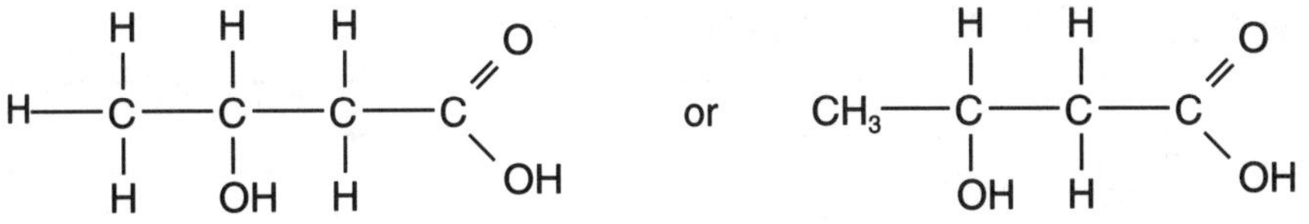

3-hydroxybutanoic acid

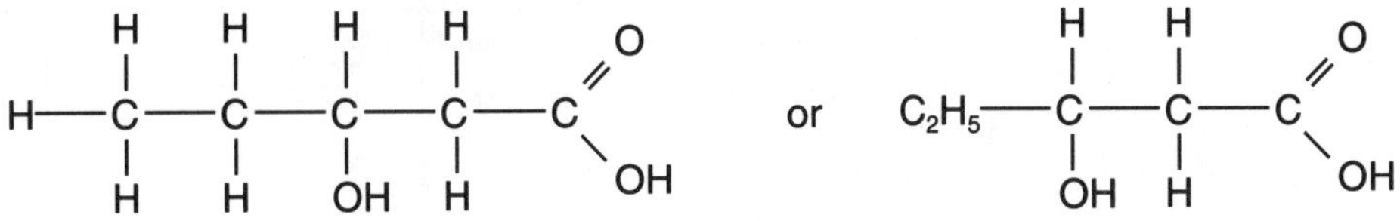

3-hydroxypentanoic acid

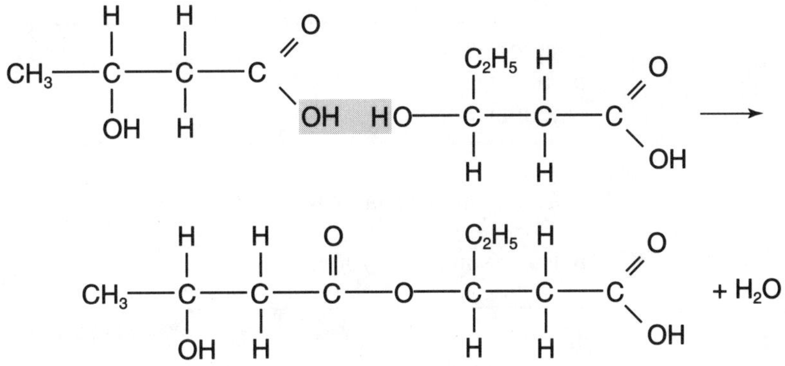

or

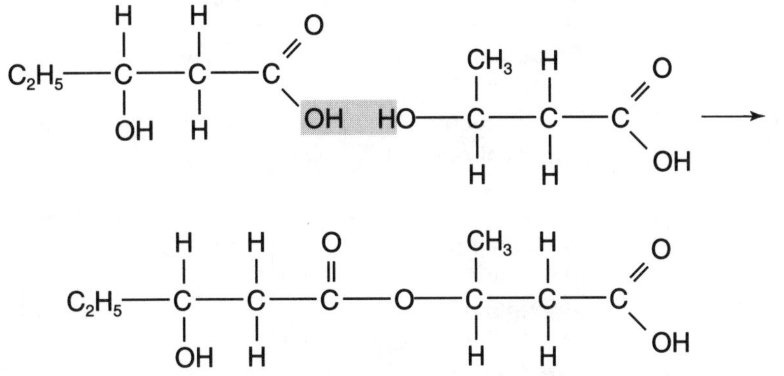

ii) The polymer will be a thermoplastic because there are no cross-links between chains.

EXERCISE 95   For a sample of 100 g:

	C	H	O
Mass/g	52.2	13.0	34.8
Molar mass/g mol^{-1}	12.0	1.0	16.0
Amount/mol	$\dfrac{52.2}{12.0} = 4.35$	$\dfrac{13.0}{1.0} = 13.0$	$\dfrac{34.8}{16.0} = 2.18$
$\dfrac{\text{Amount}}{\text{Smallest amount}}$	$\dfrac{4.35}{2.18} = 2$	$\dfrac{13.0}{2.18} = 6$	$\dfrac{2.18}{2.18} = 1$
Simplest ratio of relative amounts	2	6	1

Empirical formula of A = C_2H_6O

EXERCISE 96   **a** For a sample of 100 g:

	C	H	O
Mass/g	74.2	7.9	17.9
Molar mass/g mol^{-1}	12.0	1.0	16.0
Amount/mol	6.18	7.90	1.12
$\dfrac{\text{Amount}}{\text{Smallest amount}}$	$\dfrac{6.18}{1.12} = 5.52$	$\dfrac{7.90}{1.12} = 7.05$	$\dfrac{1.12}{1.12} = 1$
Simplest ratio of relative amounts	11	14	2

Empirical formula of A = $C_{11}H_{14}O_2$

b $M_r(C_{11}H_{14}O_2) = (11 \times 12) + (14 \times 1) + (2 \times 16) = 178$
∴ molecular formula = empirical formula = $C_{11}H_{14}O_2$

EXERCISE 97   **a** For a sample of 100 g of either compound:

	C	H	O
Mass/g	60.0	13.3	26.7
Molar mass/g mol^{-1}	12.0	1.0	16.0
Amount/mol	5.00	13.3	1.67
$\dfrac{\text{Amount}}{\text{Smallest amount}}$	$\dfrac{5.00}{1.67} = 2.99$	$\dfrac{13.3}{1.67} = 7.96$	$\dfrac{1.67}{1.67} = 1.00$
Simplest ratio of relative amounts	3	8	1

Empirical formula = **C_3H_8O**

b $M_r(C_3H_8O) = (3 \times 12.0) + (8 \times 1.0) + (1 + 16.0) = 60.0$

∴ molecular formula = empirical formula = **C_3H_8O**

c The reaction with sodium suggests that R is an alcohol. The reaction with acidified dichromate suggests that R is a primary alcohol because the product is a reducing agent (an aldehyde).

∴ R = $CH_3CH_2CH_2OH$, propan-1-ol

There are two isomers of R, propan-2-ol, $(CH_3)_2CHOH$, and methoxyethane, $CH_3CH_2OCH_3$, but propan-2-ol also reacts with sodium.

∴ Q = $CH_3CH_2OCH_3$, methoxyethane

d $CH_3CH_2CH_2OH + Na \rightarrow CH_3CH_2CH_2O^-Na^+ + \frac{1}{2}H_2$
            sodium propoxide
$CH_3CH_2CH_2OH \rightarrow CH_3CH_2CHO + 2H^+ + 2e^-$
            propanal

Since a **simplified** equation is specified, the following might be acceptable:

$CH_3CH_2CH_2OH + [O] \rightarrow CH_3CH_2CHO + H_2O$

EXERCISE 98   For a sample of 100 g:

	C	H	O
Mass/g	31.5	5.3	63.2
Molar mass/g mol^{-1}	12.0	1.0	16.0
Amount/mol	2.63	5.3	3.95
$\dfrac{\text{Amount}}{\text{Smallest amount}}$	$\dfrac{2.63}{2.63} = 1.0$	$\dfrac{5.3}{2.63} = 2.0$	$\dfrac{3.95}{2.63} = 1.5$
Simplest ratio of relative amounts	2	4	3

Empirical formula of X = **$C_2H_4O_3$**

   Since the products of the reactions of X contain only two carbon atoms, the empirical formula of X must correspond to its molecular formula $C_2H_4O_3$.

   The reaction with sodium carbonate solution to evolve CO_2 indicates that X contains at least one carboxylate group, $—CO_2H$. The reaction with PCl_5 to produce HCl indicates that X contains a hydroxyl group, $—OH$, either in a $—CO_2H$ group or on its own.

   Since the product with PCl_5 contains two chlorine atoms, X must either contain two carboxylate groups or one carboxylate and one hydroxyl group (not two hydroxyl groups, because X is acidic).

   A compound of formula $C_2H_4O_3$ cannot contain two $—CO_2H$ groups but can contain an $—OH$ and a $—CO_2H$ group.

   X must therefore be **$CH_2(OH)CO_2H$**, hydroxyethanoic acid.

   Equations for the reactions are:

$$2CH_2(OH)CO_2H + Na_2CO_3 \rightarrow 2CH_2OHCO_2Na + H_2O + CO_2$$
$$CH_2OHCO_2H + 2PCl_5 \rightarrow CH_2ClCOCl + 2POCl_3 + 2HCl$$

Two possible syntheses of 2-hydroxyethanoic acid are as follows:

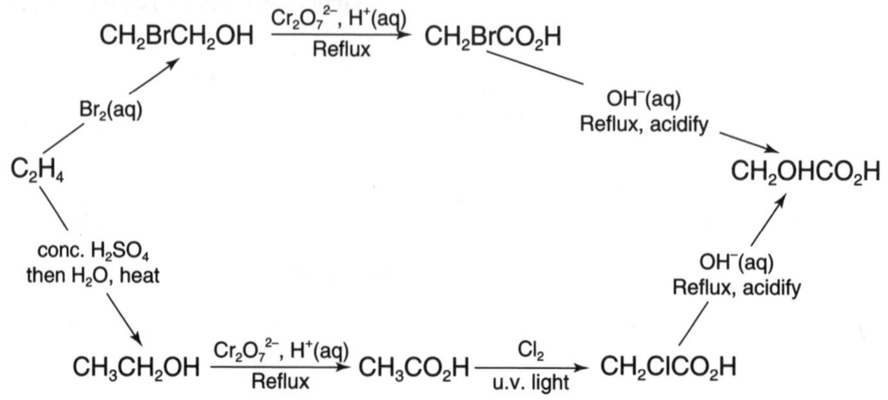

EXERCISE 99   **a** For a sample of 100 g:

	C	**H**
Mass/g	92.3	7.7
Molar mass/g mol^{-1}	12.0	1.0
Amount/mol	7.69	7.7
Simplest ratio of relative amounts	1	1

Empirical formula of A = **CH**
Substituting into the expression:
molecular formula = (empirical formula)$_n$

where $n = \dfrac{\text{relative molecular mass}}{\text{relative empirical formula mass}} = \dfrac{78}{13} = 6$

gives molecular formula = (CH)$_6$ = **C$_6$H$_6$**
The reaction with concentrated nitric acid and sulphuric acid indicates that A is benzene, C_6H_6. The nitration of benzene and the subsequent reduction show that B is nitrobenzene and C is phenylamine.

A = benzene, C$_6$H$_6$ B = nitrobenzene, C$_6$H$_5$NO$_2$ C = phenylamine, C$_6$H$_5$NH$_2$

The function of the concentrated sulphuric acid in the conversion of A into B is to react with the nitric acid and produce the nitryl cation, NO_2^+. The nitryl cation then reacts with benzene to form nitrobenzene.

$$2H_2SO_4 + HNO_3 \rightarrow H_3O^+ + NO_2^+ + 2HSO_4^-$$
$$C_6H_6 + NO_2^+ \rightarrow C_6H_5NO_2 + H^+$$

(The nitryl cation is sometimes called the nitronium ion.)
b The sodium nitrite solution reacts with the hydrochloric acid to generate nitrous acid.

$$H^+(aq) + NO_2^-(aq) \rightarrow HNO_2(aq)$$

This cooled nitrous acid then reacts with the phenylamine in acid solution to form benzenediazonium ions.

$$C_6H_5NH_3^+(aq) + HNO_2(aq) \rightarrow C_6H_5N_2^+(aq) + 2H_2O(l)$$

On warming, the solution of benzenediazonium ions gives nitrogen gas and phenol.

$$C_6H_5N_2^+(aq) + H_2O(l) \rightarrow C_6H_5OH(aq) + N_2(g) + H^+(aq)$$

This solution of phenol, when added to the original cooled mixture containing benzenediazonium ions, forms a brightly coloured yellow azo-dye.

4-hydroxyazobenzene

EXERCISE 100   For a sample of 100 g:

	C	H	N	O
Mass/g	59.4	10.9	13.9	15.8
Molar mass/g mol^{-1}	12.0	1.0	14.0	16.0
Amount/mol	4.95	10.9	0.99	0.99
$\dfrac{\text{Amount}}{\text{Smallest amount}}$	$\dfrac{4.95}{0.99} = 5.0$	$\dfrac{10.9}{0.99} = 11.0$	$\dfrac{0.99}{0.99} = 1.0$	$\dfrac{0.99}{0.99} = 1.0$
Simplest ratio of relative amounts	5	11	1	1

Empirical formula of **P** = **$C_5H_{11}NO$**
Summary of information:

$$(C_5H_{11}NO)_n \xrightarrow[\substack{\text{Distil, acidify}\\ \text{residue}}]{OH^-(aq)} \underset{Q}{C_2H_7N} \; + \; \underset{R}{C_3H_6O_2}$$

$$\underset{Q}{C_2H_7N} \xrightarrow{HNO_2} \underset{\substack{S\\ + \text{ colourless gas}}}{} \xrightarrow[\text{Heat}]{H_2SO_4} \underset{\substack{T(l)\\ (\text{Sweet smell})}}{}$$

The reaction with sodium nitrite and sulphuric acid shows that **Q** must be a primary amine. The only primary amine with molecular formula C_2H_7N is ethylamine, $C_2H_5NH_2$, and **S** must therefore be ethanol, C_2H_5OH.

The sweet-smelling liquid produced from **R** and **S** by heating with concentrated sulphuric acid must be an ester, and **R** must be a carboxylic acid. The only carboxylic acid with molecular formula $C_3H_6O_2$ is propanoic acid, $C_2H_5CO_2H$.

P must therefore be N-ethylpropanamide, $C_2H_5CONHC_2H_5$, with $n = 1$, and the ester, **T**, must be ethyl propanoate, $C_2H_5CO_2C_2H_5$.

P = N-ethylpropanamide, **$C_2H_5CONHC_2H_5$**   **Q** = ethylamine, **$C_2H_5NH_2$**
R = propanoic acid, **$C_2H_5CO_2H$**   **S** = ethanol, **C_2H_5OH**
T = ethyl propanoate, **$C_2H_5CO_2C_2H_5$**

$$\underset{P}{C_2H_5CONHC_2H_5} + OH^- \rightarrow C_2H_5CO_2^- + \underset{Q}{C_2H_5NH_2}$$

$$C_2H_5CO_2^- + H^+ \rightarrow \underset{R}{C_2H_5CO_2H}$$

$$\underset{Q}{C_2H_5NH_2} + HNO_2 \rightarrow \underset{S}{C_2H_5OH} + N_2 + H_2O$$

$$\underset{R}{C_2H_5CO_2H} + \underset{S}{C_2H_5OH} \rightarrow \underset{T}{C_2H_5CO_2C_2H_5} + H_2O$$

EXERCISE 101

a For a sample of 100 g:

	C	H	O
Mass/g	66.7	11.1	22.2
Molar mass/g mol^{-1}	12.0	1.0	16.0
Amount/mol	5.56	11.1	1.39
$\dfrac{\text{Amount}}{\text{Smallest amount}}$	$\dfrac{5.56}{1.39} = 4.0$	$\dfrac{11.1}{1.39} = 8.0$	$\dfrac{1.39}{1.39} = 1.0$
Simplest ratio of relative amounts	4	8	1

Empirical formula of X = **C_4H_8O**

b Relative mass corresponding to empirical formula, C_4H_8O:

$$M_r(C_4H_8O) = (4 \times 12) + (8 \times 1) + (1 \times 16) = 72$$

Since this is the same as the molar mass given:
molecular formula of X = **C_4H_8O**

c The positive test with 2,4-dinitrophenylhydrazine indicates the presence of a carbonyl group $C = O$.

d

EXERCISE 102

The trough at about 7×10^{-6} m in the spectrum of decane and its absence from the spectra of trichloromethane and tetrachloromethane implies that it results from the vibrations associated with the C—C bond.

$\therefore$ C—C bonds absorb radiation of wavelength **7×10^{-6} m**

The trough at about 3.5×10^{-6} m in the spectra of decane and trichloromethane, and its absence from the spectrum of tetrachloromethane, implies that it results from the vibrations associated with the C—H bond.

$\therefore$ C—H bonds absorb radiation of wavelength **3.5×10^{-6} m**.

The trough at 13.5×10^{-6} m in the spectra of trichloromethane and tetrachloromethane, and its absence from the spectrum of decane, implies that it results from the vibrations associated with the C—Cl bond.

$\therefore$ C—Cl bonds absorb radiation of wavelength **13.5×10^{-6} m**

Note that the trough at 8.4×10^{-6} m for trichloromethane is also due to absorption by C—H bonds even though it does not appear (or is displaced) for decane. This is an indication that the interpretation of IR spectra is not always simple.

EXERCISE 103   **A** is due to N—H stretching.
B is due to C—H stretching.
C is due to N—H bending.
D is due to C—C vibrations in the benzene ring.

EXERCISE 104   **B** can be identified as the infra-red spectrum of benzene from the peak between 1400 cm^{-1} and 1600 cm^{-1} due to C$\cdots$C vibrations in the benzene ring.
A can be identified as the infra-red spectrum of ethyl ethanoate from the peak between 1600 cm^{-1} and 1800 cm^{-1} due to C$=$O stretching.

EXERCISE 105   **P**: 2250 cm^{-1}, C$\equiv$N
Q: 1750 cm^{-1}, C$=$O
R: 1200 cm^{-1}, C—O

The trough at 3000 cm^{-1} is **not** broad enough to be due to an O—H group (most probably C—H stretching). Since there is no O—H, you can rule out a carboxylic acid, so X must be an ester such as:

or one of its isomers.

EXERCISE 106   The trough between about 2500 and 3500 cm^{-1} is broad enough to be due to the presence of an O—H group. (Compare this with the narrower trough caused by C—H stretching in Fig. 45.)
The trough at about 1700 cm^{-1} is due to the presence of a C$=$O group.
Figure 52 is therefore the IR spectrum for ethanoic acid.

EXERCISE 107   **a** 3340 cm^{-1}, O—H stretch
2950 cm^{-1}, C—H stretch
b Ethanol concentrations cannot be determined from the O—H vibration because of the water vapour present in both the atmosphere and the client's breath, therefore the C—H vibration is used.
c Propanone CH$_3$COCH$_3$ (acetone) also has C—H bonds that stretch at about 2950 cm^{-1}. This must therefore be eliminated from the result. Otherwise a false high reading for alcohol could be given.

EXERCISE 108   **a** The bombardment of molecules of ethanol by a high-energy electron beam not only produces ions of the parent molecule C$_2$H$_5$OH$^+$ but also fragment ions which result from some of the molecules splitting up, e.g. C$_2$H$_5^+$, CH$_2$OH$^+$, etc.
b Peak **A** ($m/e = 29$) corresponds to an ion of relative molecular mass 29; therefore, **A** = (C$_2$H$_5$)$^+$.
Peak **B** ($m/e = 31$) corresponds to an ion of relative molecular mass 31; therefore, **B** = (CH$_2$OH)$^+$.
Peak **C** ($m/e = 45$) corresponds to an ion of relative molecular mass 45; therefore, **C** = (C$_2$H$_5$O)$^+$.
Peak **D** ($m/e = 46$) corresponds to an ion of relative molecular mass 46; therefore, **D** = parent molecular ion (C$_2$H$_5$OH)$^+$.

EXERCISE 109 **a** Empirical formula determination. For a sample of 100 g:

	C	H
Mass/g	92.4	7.6
Molar mass/g mol^{-1}	12.0	1.0
Amount/mol	$\dfrac{92.4}{12} = 7.7$	$\dfrac{7.6}{1.0} = 7.6$
Simplest ratio of relative amounts	1	1

Empirical formula of X = **CH**
The M peak is 12.5 times as intense as the $(M+1)$ peak.
∴ the $(M+1)$ peak is 100/12.5 = 8% of the height of the M peak.
In a compound containing N carbon atoms, the $(M+1)$ peak is approximately N% of the height of the M peak.
∴ X contains eight carbon atoms and so:
molecular formula = **C_8H_8**
The reaction between one mol of X and one mol of Br_2 suggests that X contains one double bond. The octenes are ruled out since they have the formula C_8H_{16}. The remainder of the molecule (apart from the double bond) must be unsaturated but not reactive with bromine, i.e. it must contain a benzene ring.
∴ X must be phenylethene,

b There would be three peaks due to $CH_2{}^{79}Br_2{}^+$, $CH_2{}^{81}Br_2{}^+$ and $CH_2{}^{81}Br^{79}Br^+$. Note that peaks due to ^{13}C would be scarcely visible because the molecule has only one carbon atom.

EXERCISE 110 **a** For hydrocarbon A, the species of highest mass/charge ratio, 170, corresponds to an ion of the complete molecule being analysed. The isotopic peak due to molecular ions containing ^{13}C is too small to be visible.
  For hydrocarbon B, the species of highest mass/charge ratio 191 corresponds to an ion of the complete molecule containing the ^{13}C isotope.
b The major peaks are caused by ion fragments resulting from breaking the weaker bonds in the molecule. Associated minor peaks are caused by a variety of other ion fragments, e.g.:
  i) Major fragments as above, but with one, two or three more H atoms removed. These give smaller peaks at lower mass/charge ratio.
  ii) Major fragments containing isotopes of different mass. For a hydrocarbon, the only significant isotope is ^{13}C and this would give very small peaks at higher mass/charge ratio.
  iii) Impurities from air or residues from the previous sample. These can give small peaks anywhere along the spectrum.
c $C_8H_{17}{}^+$ (from mass/charge ratio 113 and by comparison with labelled peaks).

EXERCISE 111   The two peaks with smaller mass/charge ratio than $C_5H_{11}^+$ are due to $C_5H_{10}^+$ and $C_5H_9^+$. The one with higher mass/charge ratio at about 5% the intensity is due to $^{13}C^{12}C_4H_{11}^+$.

EXERCISE 112   a Molecular ions of mass/charge ratios 112 and 114 are due to $C_6H_5{}^{35}Cl^+$ and $C_6H_5{}^{37}Cl^+$ respectively.
b The peaks at mass/charge ratios 113 and 115 are due to some of the molecular ions from (a) containing the ^{13}C isotope.
c The fragment responsible for peak mass/charge ratio 77 is $C_6H_5^+$.

$$as\ M_r = (6 \times 12) + (5 \times 1) = 77$$

This would be produced when the parent molecule splits as shown below:

The Cl^- fragment may or may not then have electrons removed by bombardment to form a Cl^+ ion.

EXERCISE 113   a Hydrogen nuclei are shielded to a certain extent from the external magnetic field by other surrounding electrons or charge cloud.
b The extent of shielding differs for each hydrogen nucleus (proton) depending on the electron densities around it. The hydrogen atom bonded to an oxygen in an —OH group has a relatively low electron density due to the adjacent electronegative oxygen atom $-O^{\delta-}-H^{\delta+}$ and will be less shielded than a hydrogen bonded to carbon.

EXERCISE 114   a The reference compound tetramethylsilane (TMS) gives a chemical shift value of zero.
b

c i) The hydrogen atom identified as H_a in (b) above will give the highest chemical shift value, i.e. between 4 and 5, because this hydrogen has little shielding, as the electronegative oxygen next to it will reduce its surrounding electron density.
ii) The two hydrogen atoms identified as H_b are further away from the oxygen and will be more shielded than H_a. They will give the chemical shift value between 3 and 4.
iii) The three H_c hydrogen atoms are furthest from the oxygen atom and will be the most shielded and consequently give the lowest chemical shift value, between 1 and 2.
d The peak areas are in the ratio 1:2:3 (from left to right). This tells us how many of each type of hydrogen atom are present in the molecule. You may wish to summarise (a)–(d) by copying the diagram below into your notebooks.

NMR spectrum of ethanol.

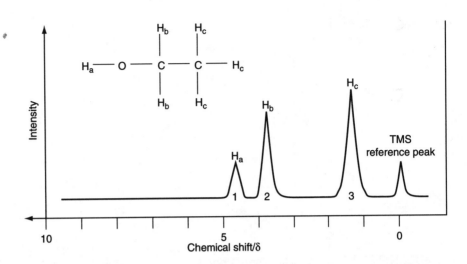

e Methoxymethane only has one type of hydrogen atom present and we should predict a shielding approximately the same as the H_b hydrogens in ethanol as they are bonded to a carbon next to an oxygen.

methoxymethane

NMR spectrum for methoxymethane.

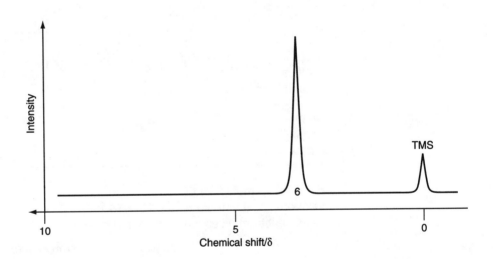

EXERCISE 115

From tables:

Type of proton	δ	Number of hydrogen atoms of this type
R—CH$_3$	0.9	9
⬡—CH$_3$	2.3	6

Area of peaks must be in the ratio 3:2 for shift values 0.9 and 2.3 respectively.

NMR spectrum for musk xylene.

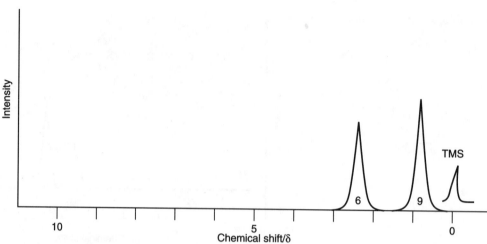

EXERCISE 116   a   Mass spectrometry.
b   Infra-red spectroscopy.
c   NMR spectroscopy.

EXERCISE 117

	C	H	O
Mass/g	48.6	8.1	43.3
Molar mass/g mol^{-1}	12.0	1.0	16.0
Amount/mol	$\dfrac{48.6}{12.0} = 4.05$	$\dfrac{8.1}{1.0} = 8.1$	$\dfrac{43.3}{16.0} = 2.71$
$\dfrac{\text{Amount}}{\text{Smallest amount}}$ = relative amount	$\dfrac{4.05}{2.71} = 1.5$	$\dfrac{8.1}{2.71} = 3.0$	$\dfrac{2.71}{2.71} = 1.0$
Simplest ratio of relative amounts	3	6	2

Empirical formula $C_3H_6O_2$
M_r of empirical formula = $(3 \times 12) + (6 \times 1) + (2 \times 16) = 74$
Mass spectrum gives a molecular ion at $M_r = 74$

$$\text{so molecular formula = empirical formula = } \mathbf{C_3H_6O_2}$$

Effervescence with sodium carbonate solution suggests a carboxylic acid – probably propanoic acid.
    Analysis of each spectrum in turn attempts to confirm this.
Mass spectrum
The peak corresponding to the second highest mass, i.e. 74, is caused by the ionised molecule of B having only ^{12}C in it. $C_3H_6O_2^+$ peak at 75 is due to a molecule ion with one ^{13}C atom in it.
    Scanning along the spectrum:

74	$C_3H_6O_2^+$	29	$C_2H_5^+$ or COH^+
57	$C_2H_5CO^+$	27	$C_2H_3^+$
45	CO_2H^+	28	$C_2H_4^+$ or CO^+

Infra-red spectrum
Working from left to right:
The very broad dip at about 3000 cm^{-1} is due to O—H (broadened due to hydrogen bonding)
The next dip at about 1700 cm^{-1} is due to C=O.
NMR spectrum
This shows three types of hydrogen atoms in the ratio 3 : 2 : 1.
   From shift values:
δ = 11.5 suggests a hydrogen atom as shown below:

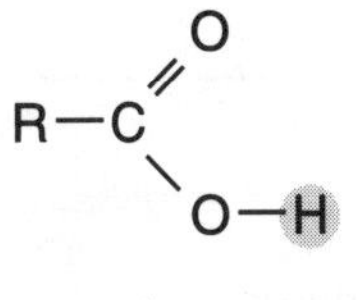

δ = 1 suggests three hydrogen atoms in a methyl group as shown below:

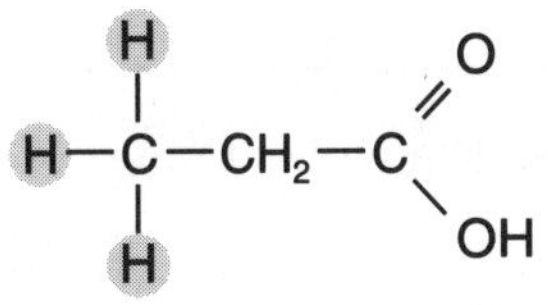

δ = 2.6 must be produced by the two hydrogen atoms adjacent to the CO$_2$H group as shown below:

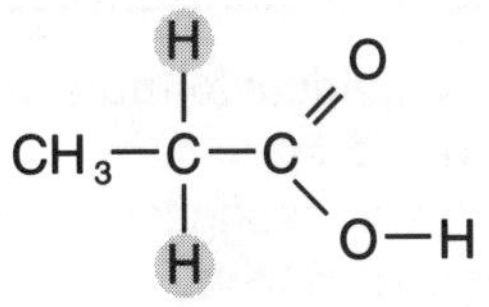

Table 19 does not contain the chemical shift value for a proton adjacent to a CO$_2$H group.

EXERCISE 118   **a** Ethanol to propanoic acid:

$$C_2H_5OH \xrightarrow[\text{Room temp.}]{PCl_5} C_2H_5Cl \xrightarrow[\text{Reflux}]{\text{KCN, ethanol}} C_2H_5CN \xrightarrow[\text{Heat}]{H^+} C_2H_5CO_2H$$
propanoic acid

Three reactions, or other possibilities – three marks for each as shown:

Any suitable reaction	(3)
Reagent	(3)
Conditions	(3)

Other routes and reagents are possible. Check with your summary reaction scheme.
   b 2-chloroethanol to aminoethanoic acid:

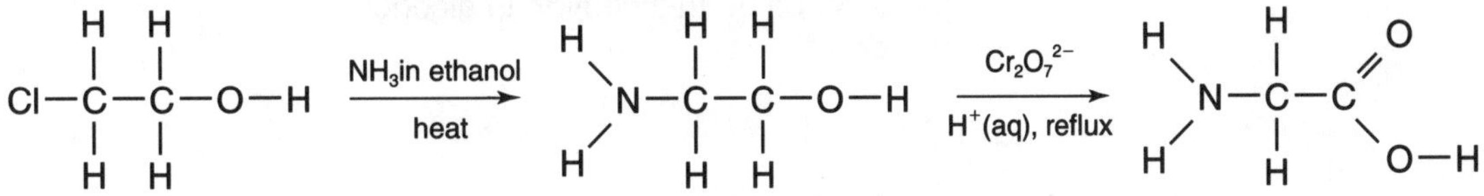

2-chloroethanol   aminoethanoic acid

Three reactions – three marks for each as follows:

Any suitable reaction	(3)
Reagent	(3)
Conditions	(3)
	(Total: 18 marks)

EXERCISE 119 a   (2)

b G is not an aldehyde.   (1)
c Carbonyl group.   (1)
d Propanone.   (1)

(Total: 5 marks)

EXERCISE 120

	C	H	O
Amount/mol	$\dfrac{66.7}{12} = 5.55$	$\dfrac{11.1}{1} = 11.1$	$\dfrac{22.2}{16} = 1.39$
Relative amount	$\dfrac{5.55}{1.39} = 4.0$	$\dfrac{11.1}{1.39} = 8.0$	$\dfrac{1.39}{1.39} = 1.0$
Ratio of relative amounts	4	8	1

The empirical formula of Z is **C_4H_8O**.   (3)
Infra-red spectrum
$1715\ cm^{-1}$ suggests $C=O$.   (2)
Mass spectrum
Peak with highest m/e ratio of 72 is due to the ionised molecule.   (1)
M_r of empirical formula = $(4 \times 12) + (8 \times 1) + (1 \times 16) = 72$.
Since mass spectrum gives a molecular ion of $M_r = 72$.
So molecular formula = empirical formula = **C_4H_6O**.   (1)
Z could be butanone, $CH_3COCH_2CH_3$, giving fragments $[CH_3COCH_2]^+$ (57)
and $[CH_3CO]^+$ (43) or butanal, $CH_3CH_2CH_2CHO$, giving fragments   (2)
$[CH_2CH_2CHO]^+$ (57) and $[CH_2CHO]^+$ (43).   (1)

(Total: 10 marks)

EXERCISE 121 a D = benzaldehyde C_6H_5CHO
E = benzoic acid $C_6H_5CO_2H$
F = phenylmethanol $C_6H_5CH_2OH$
G = sodium benzoate $C_6H_5CO_2Na$
b i) Condensation reaction to give a 2,4-dinitrophenylhydrazone.
ii) Oxidation of an aldehyde to carboxylic acid.
iii) Reduction of an aldehyde to alcohol.
c Addition.

d i)

ii) Optical/Chirality.

EXERCISE 122   **a** To each in turn add bromine in an inert solvent. Only the compound with the C=C double bond will rapidly remove the colour of the bromine by an addition reaction:

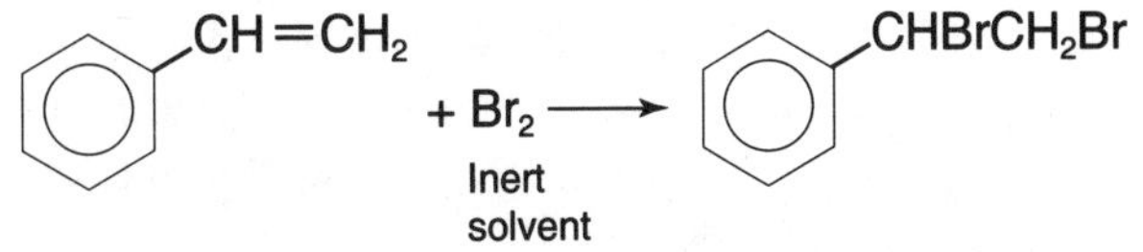

b **Either** To each in turn add $FeCl_3(aq)$ solution. Only phenol (C_6H_5OH) will give a violet colour. Equation not expected.
Or To each in turn add a little Universal Indicator. Phenol (C_6H_5OH) will be acidic at pH 4–5 whereas phenylmethanol ($C_6H_5CH_2OH$) will be neutral.

$$C_6H_5OH(aq) + H_2O(l) \rightleftharpoons C_6H_5O^-(aq) + H_3O^+(aq)$$

c To each in turn add a little ethanol and silver nitrate solution. $C_6H_5CH_2Cl$ forms a white precipitate on warming. The other compound does not. This is a substitution reaction:

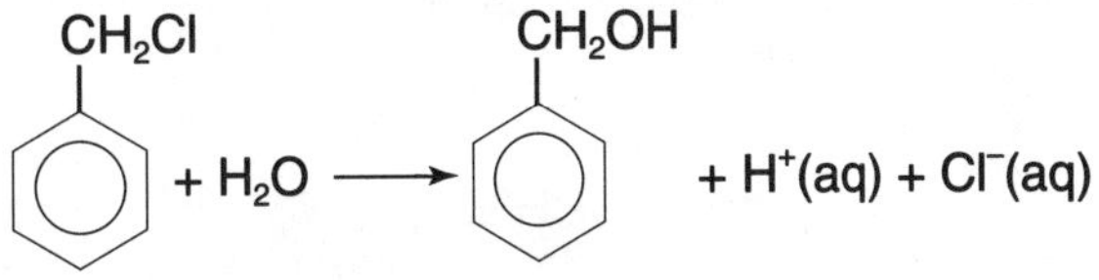

$$Ag^+(aq) + Cl^-(aq) \rightarrow AgCl(s)$$

Marks for each test as follows:	Simple test	(3)
	Observation	(3)
	Equation	(3)

(Total: 9 marks)

EXERCISE 123   **a** Bromine in an inert solvent will add to the C=C double bond.   (2)

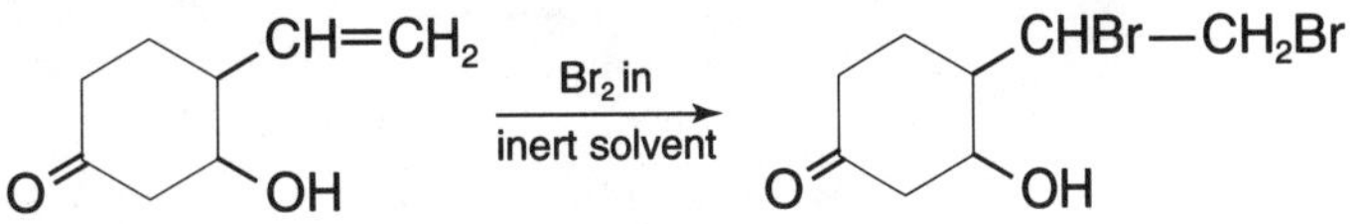

b Phosphorus pentachloride will give off fumes of HCl by reacting with the —OH group.   (2)

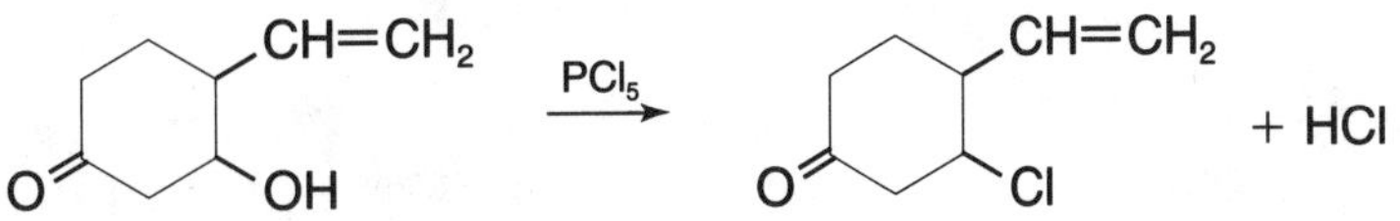

c Lithium tetrahydridoaluminate(III) will reduce the $\diagdown C=O$ (ketone group) to an alcohol.   (2)

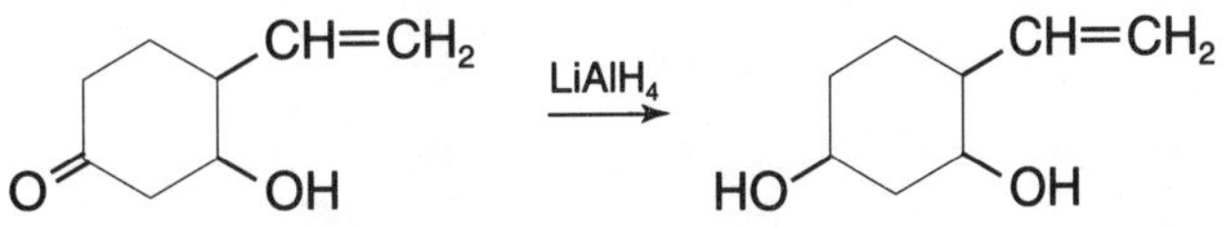

Always check that the reagent does not affect more than one functional group present. (Note: $LiAlH_4$ does not reduce C=C.)

(Total: 6 marks)

EXERCISE 124 **a** Electrophilic substitution reactions involve a replacement of one atom or group by an electrophile (electron-deficient ion or molecule). For example, the bromination of benzene. (2)

The positively charged bromine, $Br^{\delta+}$, acts as an electrophile which is attracted to the delocalised electron system in benzene to form an unstable intermediate. This is followed by the expulsion of a proton to form bromobenzene and regenerate the catalyst:

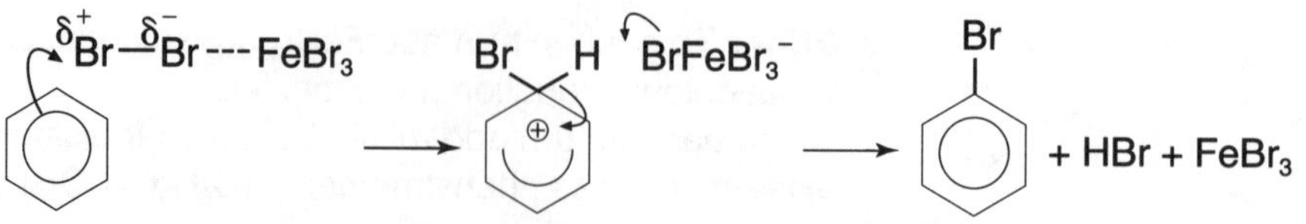

(3)

b Nucleophilic substitution reactions involve the replacement of one atom or group by a nucleophile (ions or molecules with electron-rich areas). (2)

They tend to attack positively charged areas of organic molecules. For example, nucleophilic substitution of a hydroxyl group for a halogen group in the hydrolysis of a halogenoalkane.

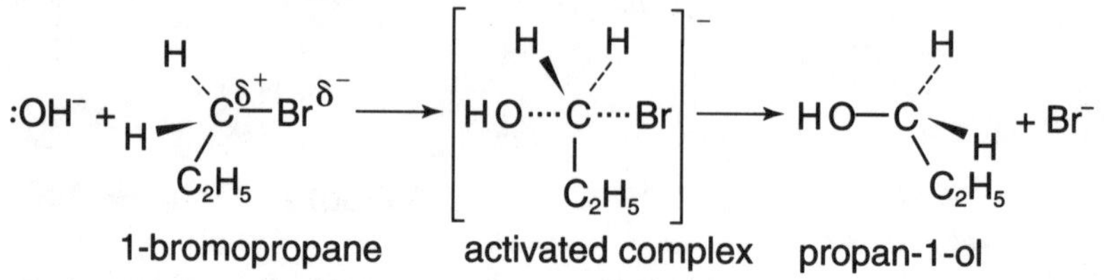

1-bromopropane   activated complex   propan-1-ol

(3)
(Total: 10 marks)

EXERCISE 125 **a** The purine bases are adenine and guanine. The pyrimidine bases are thymine and cytosine.

b

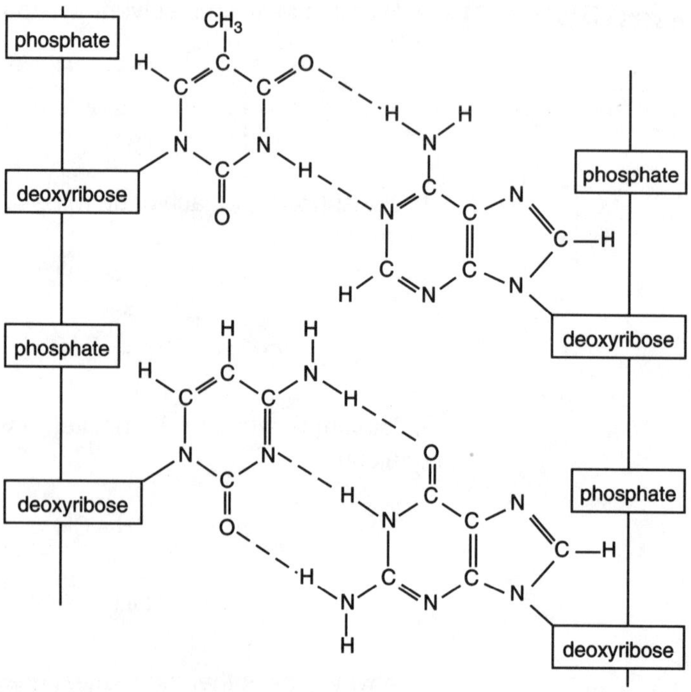

c RNA polynucleotides contain the sugar ribose (not deoxyribose as in DNA), the chains do not form a double helix, and are generally shorter.

EXERCISE 126 Amount of CO_2, $n = \dfrac{m}{M} = \dfrac{0.66 \text{ g}}{44.0 \text{ g mol}^{-1}} = 0.015 \text{ mol}$

$\therefore$ amount of C = 0.015 mol

amount of H_2O, $n = \dfrac{m}{M} = \dfrac{0.36 \text{ g}}{18.0 \text{ g mol}^{-1}} = 0.020 \text{ mol}$

$\therefore$ amount of H = 0.020 mol $\times$ 2 = 0.040 mol

	C	H
Amount/mol	0.015	0.040
$\dfrac{\text{Amount}}{\text{Smallest amount}}$	$\dfrac{0.015}{0.015} = 1.0$	$\dfrac{0.040}{0.015} = 8/3$
Simplest ratio	3	8

Empirical formula = $\mathbf{C_3H_8}$

EXERCISE 127 Amount of CO_2, $n = \dfrac{m}{M} = \dfrac{0.374 \text{ g}}{44.0 \text{ g mol}^{-1}} = 8.50 \times 10^{-3} \text{ mol}$

$\therefore$ amount of C = 8.50×10^{-3} mol

Amount of H_2O, $n = \dfrac{m}{M} = \dfrac{0.154 \text{ g}}{18.0 \text{ g mol}^{-1}} = 8.56 \times 10^{-3} \text{ mol}$

$\therefore$ amount of H = 2 $\times$ amount of water = 0.0171 mol
Mass of carbon, $m = nM = 8.50 \times 10^{-3}$ mol $\times$ 12.0 g mol^{-1} = 0.102 g
Mass of hydrogen, $m = nM = 0.0171$ mol $\times$ 1.0 g mol^{-1} = 0.0171 g
Mass of oxygen = mass of sample $-$ (mass of C + mass of H)
$\qquad\qquad$ = 0.146 g $-$ (0.102 + 0.0171) g = 0.027 g

	C	H	O
Mass/g			0.027
Molar mass/g mol^{-1}			16.0
Amount/mol	8.50×10^{-3}	0.0171	1.7×10^{-3}
$\dfrac{\text{Amount}}{\text{Smallest amount}}$	$\dfrac{8.50 \times 10^{-3}}{1.7 \times 10^{-3}} = 5.0$	$\dfrac{0.0171}{1.7 \times 10^{-3}} = 10$	$\dfrac{1.7 \times 10^{-3}}{1.7 \times 10^{-3}} = 1.0$
Simplest ratio of relative amounts	5	10	1

Empirical formula = $\mathbf{C_5H_{10}O}$

EXERCISE 128   Amount of CO_2, $n = \dfrac{m}{M} = \dfrac{0.3771 \text{ g}}{44.0 \text{ g mol}^{-1}} = 8.57 \times 10^{-3}$ mol

∴ amount of C = 8.57×10^{-3} mol

Amount of H_2O, $n = \dfrac{m}{M} = \dfrac{0.0643 \text{ g}}{18.0 \text{ g mol}^{-1}} = 3.57 \times 10^{-3}$ mol

∴ amount of H = $2 \times 3.57 \times 10^{-3}$ mol = 7.14×10^{-3} mol

Amount of AgBr, $n = \dfrac{m}{M} = \dfrac{0.2685 \text{ g}}{187.8 \text{ g mol}^{-1}} = 1.43 \times 10^{-3}$ mol

∴ amount of Br = 1.43×10^{-3} mol

	C	**H**	**Br**
Amount/mol	8.57×10^{-3}	7.14×10^{-3}	1.43×10^{-3}
$\dfrac{\text{Amount}}{\text{Smallest amount}}$	$\dfrac{8.57}{1.43} = 5.99$	$\dfrac{7.14}{1.43} = 4.99$	$\dfrac{1.43}{1.43} = 1.00$
Simplest ratio of relative amounts	6	5	1

Empirical formula = $\mathbf{C_6H_5Br}$

EXERCISE 129   **a** We would predict two resonance peaks.
b We would predict a ratio of peak area 1 : 2.

EXERCISE 130   The H_a protons have two adjacent H_b protons which are responsible for the 1 : 2 : 1 triplet ($\delta \approx 1.0$).
   The H_c proton also has two adjacent H_b protons and also gives a 1 : 2 : 1 triplet ($\delta \approx 5.0$).
   The pattern caused by the H_b protons looks rather complex and comprises eight peaks (δ 3–4). This is because it has two different types of hydrogen atoms adjacent, i.e. three H_a protons and one H_c proton. The three H_a protons produce a 1 : 3 : 3 : 1 splitting pattern, whilst the H_c proton produces a 1 : 1 doublet. Since the H_a and H_c protons are chemically different they will operate independently. Thus the H_a protons would cause 1 : 3 : 3 : 1 quartet and each peak will be split into a doublet by the H_c proton, aligning with or against the applied field. Hence the eight peaks produced by H_b protons.

EXERCISE 131   **a** i) The H_c proton produces a single peak instead of a triplet.
        ii) The H_b protons produce the normal 1 : 3 : 3 : 1 pattern.
   b i) The exchange of the H_c proton occurs so rapidly that the H_b protons only experience the averaged effect of all protons exchanging on the oxygen atom with H_c. This means that the H_b protons do not experience local fields caused by the H_c proton with its two possible orientations.
        ii) ^2_1H does not absorb in this region of the spectrum, so as H_c is replaced by ^2_1H from the $^2_1H_2O(D_2O)$, so the H_c peak at 5.7 disappears completely.

EXERCISE 132

Wave motion	Wave length (cm)	Array
Microwaves	1–10	Crystal model made from 5 cm spheres
Water waves on a pond	10–30	Row of fence posts or railings
Light (visible)	10^{-4}	Piece of closely-woven cloth
X-rays	10^{-6}–10^{-10}	Crystal

EXERCISE 133

a

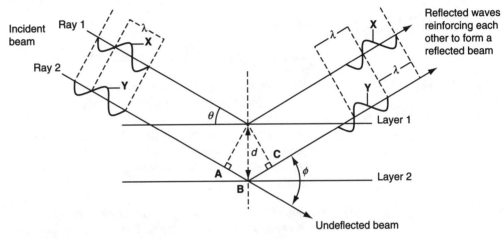

b The distances travelled by rays 1 and 2 must differ by a whole number of wavelengths, i.e.

$$AB + BC = n\lambda$$

but $\qquad AB = BC = d \sin\theta$

or $\qquad 2d \sin\theta = n\lambda \quad$ (This is the Bragg equation.)

c $\phi = 2\theta$

This is why 2θ appears in some diagrams similar to Fig. 89, rather than θ.

d Several peaks appear on each trace because reinforcement of the reflected rays occurs for different values of n. These decrease in intensity as n increases from 1, although this is not shown clearly in Fig. 89.

e Three different traces are obtained for three different sets of planes as the crystal is rotated. These planes are shown in Fig. 87, page 159.

EXERCISE 134

$n\lambda = 2d \sin\theta$

Assume $n = 1$ for a fairly strong reflected beam

$\therefore 0.0583$ nm $= 2d \sin 9°$

$\qquad\qquad\qquad = 2d \times 0.156$

$\therefore d = \dfrac{0.0583 \text{ nm}}{2 \times 0.156} = \textbf{0.187 nm}$

EXERCISE 135

$\theta = \frac{1}{2}\phi$ (see Fig. 88, page 160).

$\quad = \frac{1}{2}(47°30')$

$\quad = 23°45'$

$n\lambda = 2d \sin\theta$ or $d = \dfrac{n\lambda}{2 \sin\theta}$

$\therefore d = \dfrac{n \times 0.0635 \text{ nm}}{2 \sin 23°45'} = \dfrac{n \times 0.0635}{2 \times 0.4027}$ nm $= n \times 0.0788$ nm

If $n = 1$ then $d = \textbf{0.0788 nm}$

If $n = 2$ then $d = \textbf{0.158 nm}$

If $n = 3$ then $d = \textbf{0.236 nm}$

The first two values are rather small, compared with the diameters of most atoms and ions, which makes the third most likely. The best method of distinguishing them, however, would be to look for other reflected beams at angles obtained by substituting these values for d into the Bragg equation using different values of n. For instance, if 0.236 nm is the correct value, then there should be a brighter beam ($n = 2$) at $\theta = 15\frac{1}{2}°$ and a still brighter beam ($n = 1$) at $\theta = 8°$. If 0.158 nm is the correct value, then there should be a brighter beam ($n = 1$) at $\theta = 11\frac{1}{2}°$ and a weaker beam ($n = 3$) at $\theta = 37°$.

EXERCISE 136

$2d \sin \theta = n\lambda$

$\therefore 2 \times 0.198 \text{ nm} \times \sin 12° = 1 \times \lambda$

i.e. $\lambda = 2 \times 0.198 \text{ nm} \times 0.208 = \textbf{0.0824 nm}$

EXERCISE 137

a The crystal is rotated so that various sets of planes come into position at the correct angle to give reinforcing reflected beams.

b The spots (or rings) in X-ray diffraction patterns vary in intensity because atoms with large numbers of electrons (dense electron clouds) diffract X-rays more strongly than those with small numbers of electrons.

c Hydrogen atoms have low-density electron clouds and therefore diffract X-rays very weakly.

EXERCISE 138

a Covalent bonding. The contours show a significant electron density in the region between neighbouring atoms. This indicates the electron sharing which is characteristic of covalent bonding.

b A, H and I are oxygen atoms – they have greater electron density than B, C, D, E, F, G, J and K which are therefore carbon atoms.

c Hydrogen atoms cannot be located precisely, but their presence is indicated by the outward projections from the contour of lowest electron density.